Elliptical Mirrors

Applications in microscopy

SERIES EDITOR

Professor Rajpal S Sirohi Consultant Scientist

About the Editor

Rajpal S Sirohi is currently working as a faculty member in the Department of Physics, Alabama A&M University, Huntsville, Alabama (USA). Prior to this, he was a consultant scientist at the Indian Institute of Science Bangalore, and before that he was chair professor in the Department of Physics, Tezpur University, Assam. During 2000–11, he was academic administrator, being vice chancellor to a couple of universities and the director of the Indian Institute of Technology Delhi. He is the recipient of many international and national awards and the author of more than 400 papers. Dr Sirohi is involved with research concerning optical metrology, optical instrumentation, holography, and speckle phenomenon.

About the series

Optics, photonics and optoelectronics are enabling technologies in many branches of science, engineering, medicine and agriculture. These technologies have reshaped our outlook, our way of interaction with each other and brought people closer. They help us to understand many phenomena better and provide a deeper insight in the functioning of nature. Further, these technologies themselves are evolving at a rapid rate. Their applications encompass very large spatial scales from nanometers to astronomical and a very large temporal range from picoseconds to billions of years. The series on the advances on optics, photonics and optoelectronics aims at covering topics that are of interest to both academia and industry. Some of the topics that the books in the series will cover include bio-photonics and medical imaging, devices, electromagnetics, fiber optics, information storage, instrumentation, light sources, CCD and CMOS imagers, metamaterials, optical metrology, optical networks, photovoltaics, freeform optics and its evaluation, singular optics, cryptography and sensors.

About IOP ebooks

The authors are encouraged to take advantage of the features made possible by electronic publication to enhance the reader experience through the use of colour, animation and video, and incorporating supplementary files in their work.

Do you have an idea of a book you'd like to explore?

For further information and details of submitting book proposals see **iopscience.org/books** or contact Ashley Gasque on **Ashley.gasque@iop.org**.

Elliptical Mirrors

Applications in microscopy

Jian Liu
Harbin Institute of Technology, China

IOP Publishing, Bristol, UK

ISBN   978-0-7503-1629-3 (ebook)
ISBN   978-0-7503-1627-9 (print)
ISBN   978-0-7503-1628-6 (mobi)

DOI   10.1088/978-0-7503-1629-3

Version: 20181001

IOP Expanding Physics
ISSN 2053-2563 (online)
ISSN 2054-7315 (print)

British Library Cataloguing-in-Publication Data: A catalogue record for this book is available from the British Library.

Published by IOP Publishing, wholly owned by The Institute of Physics, London

IOP Publishing, Temple Circus, Temple Way, Bristol, BS1 6HG, UK

US Office: IOP Publishing, Inc., 190 North Independence Mall West, Suite 601, Philadelphia, PA 19106, USA

Cover image: Cut from a blue fiber optic cable. Credit: nikkytok/shutterstock.

We are grateful for the support from the Excellent Youth Foundation of Heilongjiang Scientific Committee No. JC2017013, Equipment pre-research field fund No. 6140923020102, and Key Laboratory of Micro-systems and Micro-structures Manufacturing of the Ministry of Education.

Contents

Series preface

Optics, photonics and optoelectronics are enabling technologies in many branches of science, engineering, medicine and agriculture. These technologies have reshaped our outlook, our way of interaction with each other and brought people closer. They help us to understand many natural phenomena better and provide a deeper insight into the functioning of nature. Further, these technologies themselves are evolving at a rapid rate. Their applications encompass very large spatial scales from nanometers to astronomical and a very large temporal range from femtoseconds to billions of years. The series on advances on optics, photonics and optoelectronics aims to cover topics that are of interest to both academia and industry. Some of the topics that the books in the series will cover are bio-photonics and medical imaging, devices, electromagnetics, fiber optics, information storage, instrumentation, light sources, CCD and CMOS imagers, metamaterials, optical metrology, optical networks, photovoltaics, freeform optics and its evaluation, singular optics, cryptography and sensors.

Short CV

Professor Rajpal S Sirohi has been working in optics since 1965 in various capacities. He was professor of at the Indian Institute of Technology Madras (India) since 1979. He was visiting professor at the National University of Singapore and E.P.F.L, Lausanne (Switzerland). He worked as an Alexander von Humboldt (AvH) fellow at P.T.B. Braunschweig (Germany) and an AvH Awardee at Oldenburg University (Germany). He was ICTP consultant/visiting scientist to the University of Malaya (Malaysia) and Namibia University (Namibia).

During 2000–13, he was an academic administrator, being the director of the Indian Institute of Technology Delhi and vice chancellor of several universities.

He has been on the editorial board of several journals. He was associate editor for Optical Engineering during 2000–13. Since 2013, he has been the senior editor of Optical Engineering.

Professor Sirohi has received many national and international awards. His notable awards include the Galillee Gallileo Award of International Commission of Optics, Gabor and Vikram Awards by SPIE and Padma Shri by the Government of India. He is a fellow of INAE, NASI, OSA, SPIE, ISoI, OSI.

Professor Sirohi has authored more than 400 papers. His research areas are optical metrology, instrumentation, holography and speckle phenomenon.

Presently, he is a faculty member in the Department of Physics, Alabama A&M University, Huntsville, Alabama (USA).

Preface

High NA optical elements are basically required to setup a microscope as specimen details generally propagates as high frequency signals. Since 2010, we have investigated elliptical mirrors (EMs). When this book is published, this will hopefully be the third solution for far field imaging after parabolic mirrors and objective lenses, where apodization factors were established in 1921 and 1959, respectively. Apodization factor is definitely a key to enable a new focusing theory concerned with high NA imaging. Nevertheless, it is very hard at the beginning to explain how beam energy gets transformed on mirror surfaces, particularly when mirror function is an aspheric one. Fortunately, the apodization factor of EMs was finally figured out and published in 2012, and thus my group can go through the focusing properties of EMs in the following years.

After the past investigations, there are now a total of four original formulas (Apodization factor in chapter 2, Cylindrical-vector focusing in chapter 3, Vectorial focusing in chapter 4, and Scalar focusing in chapter 5). We dreamed to have a book until EMs were successfully applied in industrial metrology (chapter 8) and TIRF microscope with shadowless illumination and adjustable penetration depth (chapter 9). Despite current theoretical points potentially being debatable, the authors agreed to finish the book and make research continue going in an open way.

From the author's point of view, short wave, super-wide spectrum, high NA focusing and x-ray topography will be part of the trend of microscope development. EMs offer a new way, alongside objective lenses, to setup optical microscopes, and it possibly opens a door for high NA and achromatic imaging. In addition, this book will benefit researchers, engineers and students with 3D transfer function theory, focusing theory, and share experiences on mirror system design.

We thank Prof. Tony Wilson from the University of Oxford (UK) for initial discussions on the apodization factors, and Prof. Konyakhin Igor form the National Research University of Information Technologies, Mechanics and Optics (Saint Petersburg, Russia) for providing the scanned papers published by V S Ignatovsky one century ago, which essentially inspired us with a brief understanding on mirror energy transformation.

J Liu
Harbin, China
July, 2018

Acknowledgement

This work is funded by the National Natural Science Foundation of China (51275121) and Heilongjiang Science Foundation for Young Talents of China (JC2017013), Key Laboratory of Micro-system and Micro-structures Manufacturing of Ministry of Education, Harbin Institute of Technology, Harbin, 150080, China.

Editor biography

Jian Liu

Jian Liu is professor and vice dean of the School of Electrical Engineering and Automation, Harbin Institute of Technology, China, and an honorary professor at the University of Nottingham, UK. His academic interests lie in the theories and implementations of optical microscopes, in particular the development of optical microscopes, applied optics and optical metrology. He is also a council member of the China Optical Society for Engineering and China Instrument and Control Society, a member of ISO/TC213 and China SAC/TC240 as well, which are both served for geometrical products specifications, and a board member of the Journal of Microscopy, Surface Topography: Metrology & Properties, Optics Communications, Nanomanufacturing and Metrology and Applied Optics (China).

List of contributors

Min Ai
The University of British Columbia, Vancouver, Canada

Shan Gao
Harbin Institute of Technology, Harbin, China

Yong Li
Harbin Institute of Technology, Harbin, China

Mengzhou Li
Harbin Institute of Technology, Harbin, China

Qiang Li
Harbin Institute of Technology, Harbin, China

Jian Liu
Harbin Institute of Technology, Harbin, China

Chenguang Liu
Harbin Institute of Technology, Harbin, China

Jiubin Tan
Harbin Institute of Technology, Harbin, China

Yuhang Wang
Harbin Institute of Technology, Harbin, China

Chao Wang
Huawei Technologies Co Ltd, Shenzen, China

Tong Wang
China Academy of Space Technology, Beijing, China

He Zhang
Harbin Institute of Technology, Harbin, China

Cien Zhong
China Helicopter Research and Development Institute, Jingdezhen, China

Chapter 1

Research and application of reflective microscopy

Chenguang Liu, Chao Wang, Jian Liu and Jiubin Tan

1.1 Introduction

The reflective imaging system has unique advantages compared with the refractive imaging system. The reflective imaging system is free from chromatic aberration, neither the axial chromatic aberration nor the lateral chromatic aberration. In theory, the reflective microscopy can be applied to any spectral range as long as a corresponding reflective film exists in that band, but the refractive system is limited by the transmittance and refractive index of material in different spectra. Moreover, the aberration caused by a reflective spherical surface is less than that caused by a refractive lens [1], and the reflective imaging system is simple in structure, easy to make, and easy to be expanded to a large aperture size. Accordingly, the reflective imaging system is widely used in such fields, which requires a large aperture scale as the astronomical telescope and space telescope, as well as the special microscopy field that has a strict requirement for chromatic aberration.

1.2 Current situation of research on reflective microscopy

The reflective microscopy is free from limitation of materials, can be used in the range of x-ray to infrared wave bands, and has the incomparable advantage of wide spectrum imaging over the transmissive microscopy imaging.

The objective structure consisting of two reflective spherical surfaces that are concentric to each other is commonly called the Schwarzschild objective (SO) [2]. The SO structure is the first reflective objective applied in the field of microscopic study, and any further development of the reflective microscopy is based on the SO structure. The SO model was originally proposed as a solution to the imaging of an object at infinity, which is made of two spherical mirrors. With the development of demand and the deepening of research, Shealy, Hoover, Artyukov *et al* proposed a Schwarzschild objective model for finite conjugate imaging [1, 3–5]. That model was widely used in the field of x-ray study, and is still made of two spherical mirrors. With the emergence and development of the aspherical surface, to obtain better image quality and more flexible structure, Head proposed an aspherical reflective objective model [6], both mirrors of which are aspherical, and the spherical SO model became a special case of reflective objective model with an aspherical coefficient of 0.

The infinite conjugate Schwarzschild objective was first proposed in a paper by the German scientist Karl Schwarzschild in 1905. This objective, which is different from other refractive telescopes, was used first for a telescopic system, and played an important role in the astronomical observation research because it showed good imaging performance in wide spectra. Then this structure was used for the micro-scopy field, and played a unique role in fields such as spectro-microscope and x-ray.

The infinite conjugate Schwarzschild objective is composed of a convex spherical mirror and a concave spherical mirror, and its schematic diagram is as shown in figure 1.1. Element 1 and Element 2 are respectively the convex spherical mirror and the convex spherical mirror, which are called the secondary mirror and the primary mirror. Position N represents the imaging position, and position c represents the center sphere of Element 1 and Element 2 and is also the diaphragm position. d is the distance between two mirrors, b is the distance from the primary mirror to the image surface, f is the focal length, and y_1 and y_2 respectively represent the height of light rays arriving at both mirrors.

In the infinite conjugate Schwarzschild system, if the two spherical mirrors are concentric to each other, the aperture diaphragm is at the common center of sphere c, and the spherical aberration, coma and astigmatism are small. According to the

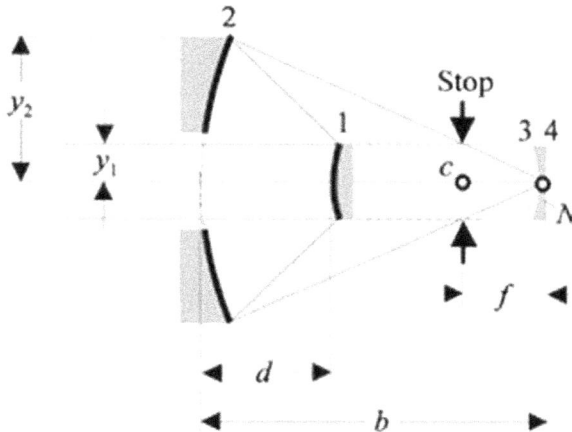

Figure 1.1. A schematic diagram of infinite conjugate Schwarzschild objective.

approximation of the Taylor series expansion of trigonometric function, the parameters need to meet the following conditions:

$$\begin{cases} d = 2f \\ b = (\sqrt{5} + 2)f \\ R_1 = (\sqrt{5} - 1)f \\ R_2 = (\sqrt{5} + 1)f \\ y_2 = (\sqrt{5} + 2)f \end{cases} \tag{1.1}$$

where R_1 and R_2 are respectively the radii of curvature of the secondary mirror and the primary mirror, and f is the focal length of the system. Formula (1.1) is the infinite imaging Schwarzschild model, which preferably eliminates the spherical aberration of the system [2]. If the focal length f is given, various parameters of the system can be obtained, including the radii of curvature of the primary mirror and the secondary mirror, object distance, image distance and the distance between two mirrors.

In 1934, Bureau and Swings tried to solve the derived problem of how to focus the system on a finite distance [6]. A real breakthrough in adjusting the object distance to a finite distance came in 1947. Burch inverted a Schwarzschild–Chretien telescope as the basis of a reflective objective, experimentally rectified it to a finite working distance, and corrected the theoretical resolution of tube with finite length [7]. In 1989, Shealy *et al* deduced the finite conjugate Schwarzschild objective model [5], and proved that there are two special object plane positions; when the object plane is at these positions, the system is aplanatic, as shown in figure 1.2. At this moment, the distance between the object plane and the center of the reflective spherical surface is:

$$Z_0 = \frac{R_1 R_2}{R_1 - R_2 \pm \sqrt{R_1 R_2}}, \quad R_1 > R_2. \tag{1.2}$$

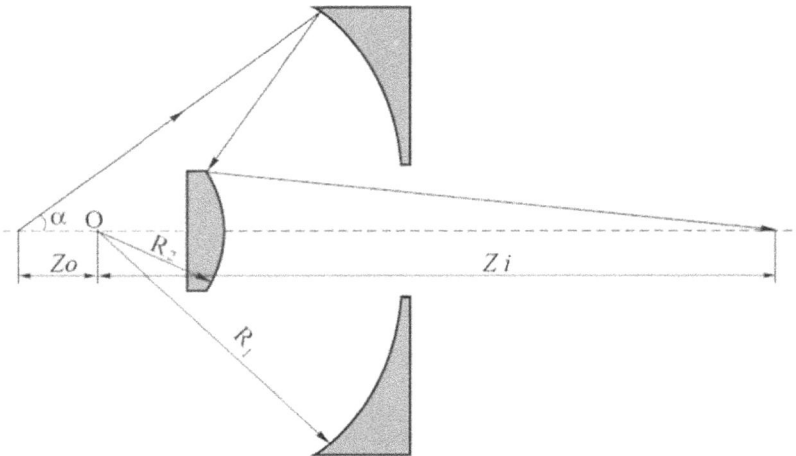

Figure 1.2. Schematic diagram of finite conjugate Schwarzschild objective.

The magnification of the corresponding objective is:

$$M_0 = -\frac{R_1 - R_2 \pm \sqrt{R_1 R_2}}{R_1 - R_2 \mp \sqrt{R_1 R_2}}, \quad R_1 > R_2 \tag{1.3}$$

where, R_1 and R_2 respectively represent the radii of curvature of the primary mirror and the secondary mirror, and M_0 is a negative value and means that the image is an inverted one.

Under the paraxial approximation, the Schwarzschild objective can be represented by a thin lens model, whose focal length is:

$$f = -\frac{R_1 R_2}{2(R_1 - R_2)}, \quad R_1 > R_2. \tag{1.4}$$

Formulas (1.2)–(1.4) involve the finite conjugate Schwarzschild objective model, and when the finite conjugate Schwarzschild objective is designed, if the radii of curvature R_1 and R_2 of the primary mirror and the secondary mirror are given, all the structural parameters of the Schwarzschild objective can be obtained by formulas (1.3) and (1.4). The design method of the Schwarzschild objective is simple but has a deficiency in principle, that is the obscuration is not taken into consideration.

The aspherical optics can better reduce the number of lenses, improve image quality and optimize system structure than the spherical optics. Therefore, with the continuous improvement in ultra-precision processing technology and optical detection technology, the aspherical surface has been widely used in the system that requires a high level of image quality, and the application of aspherical surfaces is an irresistible trend of development of the optical system.

Since the early research of Karl Schwarzschild, all of the traditional research methods have been based on the approximation of Taylor series expansion of trigonometric function [3], and have involved the spherical mirror. In 1957, Head proposed the design of an aplanatic objective without approximation, and deduced an accurate aspherical analytical model under the Abbe sine conditions and the axis astigmatism conditions in a polar coordinate system [5].

Head's objective model is shown in figure 1.3, where the axial distance from the plane where the object is located (i.e. the object plane) to the primary mirror is ρ_0,

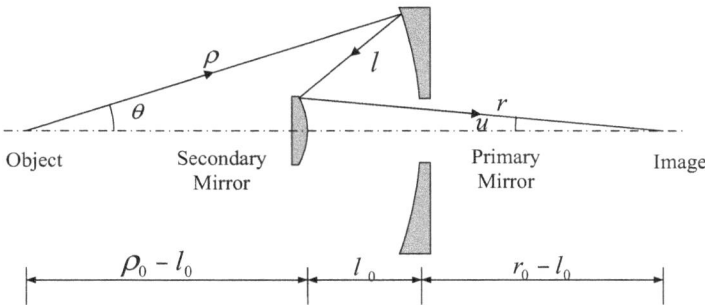

Figure 1.3. A schematic diagram of an aspherical reflective objective.

the distance between the primary mirror and the secondary mirror is l_0, and the axial distance from the secondary mirror to the image surface is r_0.

Head used the following transformation during the deduction:

$$\gamma = \cos\theta + \sqrt{m^2 - \sin^2\theta}. \tag{1.5}$$

That is:

$$\cos\theta = \frac{\gamma^2 - m^2 + 1}{2\gamma} \tag{1.6}$$

where, m represents the magnification and θ is the angle between the object space emergent ray and the optical axis.

Ultimately, Head obtained two expressions of aspherical mirror in polar coordinates, where the expression of aspherical equation of the primary mirror is as follows:

$$
\begin{aligned}
\frac{l_0}{\rho} = {} & \frac{1+\kappa}{2\kappa} + \frac{1-\kappa}{2\kappa}\cos\theta + \left[\frac{l_0}{\rho_0} - \frac{1}{\kappa}\right]\left[\frac{\gamma}{1+m}\right]^{-1}\left[\frac{\gamma - (1-m)}{2m}\right]^{\alpha} \\
& \times \left[\frac{\gamma - (m-1)}{2}\right]^{\beta}\left[\frac{\kappa+1}{2(m+1)}\gamma - \frac{\kappa-1}{2}\right]^{2-\alpha-\beta}
\end{aligned} \tag{1.7}
$$

where, $\kappa = (\rho_0 + \rho_0)/\lambda_0$, $\alpha = \mu\kappa/(\mu\kappa - 1)$, $\beta = \mu/(\mu - \kappa)$.

The expression of the aspherical equation of the secondary mirror in polar coordinates is as follows:

$$
\begin{aligned}
\frac{l_0}{r} = {} & \frac{1+\kappa}{2\kappa} + \frac{1-\kappa}{2\kappa}\cos u + \left[\frac{l_0}{r_0} - \frac{1}{\kappa}\right]\left[\frac{\delta}{1+M}\right]^{-1}\left[\frac{\delta - (1-M)}{2M}\right]^{\alpha'} \\
& \times \left[\frac{\delta - (M-1)}{2}\right]^{\beta'}\left[\frac{\kappa+1}{2(M+1)}\delta - \frac{\kappa-1}{2}\right]^{2-\alpha'-\beta'}
\end{aligned} \tag{1.8}
$$

where $\alpha' = M\kappa/(M\kappa - 1)$, $\beta' = M/(M - \kappa)$, $M = 1/\mu$, $\delta = M\gamma = \cos u + \sqrt{M^2 - \sin^2 u}$.

Although the expressions of the primary and the secondary mirror aspherical equations in polar coordinates are given in formulas (1.7) and (1.8), it can be seen that various parameters of the aspherical surface are interwoven, a specific parameter cannot be solved, and the expressions cannot be applied to the practical design [47].

All of the above coaxial reflective objective models have a common deficiency in design; the effect caused by the obscuration of the secondary mirror is not taken into consideration. The secondary mirror blocks some part of light from the object plane, and only the light at the edge contributes to the imaging. The existence of the obscuration not only reduces the brightness of the image, but also lowers the image quality, particularly in the low frequency part. According to [45], curves of optical transfer function corresponding to different shading ratios are plotted in figure 1.4.

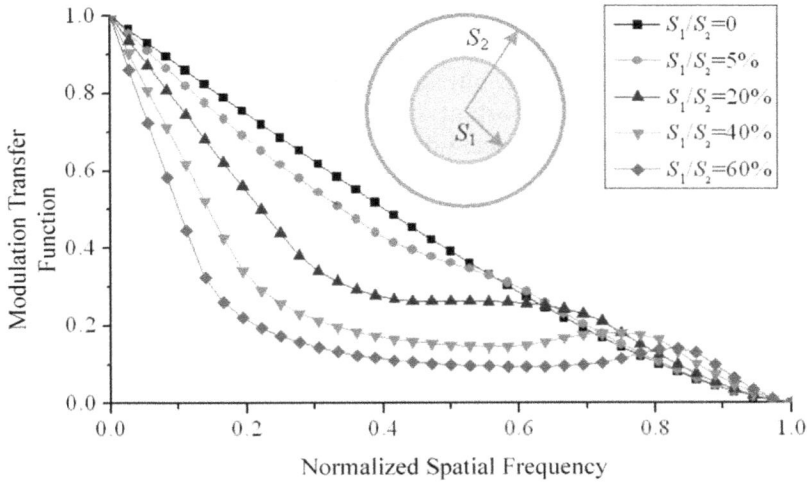

Figure 1.4. The effect of obscuration on modulation transfer function [45].

As can be seen from figure 1.4, when the shading ratio is 0, the modulation transfer function curve of the system is the highest one, which reaches the diffraction limit, and therefore, the best image can be obtained. However, in the practical application, because of the aperture of the secondary mirror, some part of light will certainly be blocked, so the shading ratio cannot reach 0. With the increase in the shading ratio, the modulation transfer function curve goes down continuously, and when the obscuration ratio reaches 20%, the curve begins to go down rapidly. This causes the reduction in image contrast, particularly in the low frequency part. The higher the obscuration ratio, the more the contrast reduces, and the greater the influence on the image quality. Therefore, to provide the reflective imaging system with a better image quality, the size of the secondary mirror shall be minimized to lower the obscuration ratio.

It is worth mentioning that the Schwarzschild objective is made of two spherical mirrors concentric to each other. Previous research on the Schwarzschild objective mainly focused on the imaging characteristics of spherical reflection, and so did the calculation. However, the only one variable parameter in the spherical surface design is the radius, and it is difficult to satisfactorily improve the image resolution and lower the obscuration ratio of the system simultaneously by the design of spherical radius parameters. Therefore, the introduction of the quadric conic surface and even the high-order aspherical surface into the system is taken into consideration. Because of the introduction of the aspherical coefficient, the design freedom can be increased to meet the requirements of imaging characteristics and shading ratio at the same time.

1.3 The current situation of application of reflective microscopy

The reflective microscopy, because of its wide applicable spectrum, has been widely used for the special microscopic imaging to which a conventional microscope is not applicable, for example, x-ray imaging, EUV/UV lithography, UV band micro–nano processing, and near-infrared imaging. It is not only a supplement to the transmissive microscopy imaging, but also an indispensable imaging technology.

X-ray microscopy is an important field of reflective microscopy application. X-ray microscopy is an imaging technology whose resolution is much higher than that of the optical microscope, with the fundamental principle that specific signals generated by the interaction between x-rays and the material are used to restore the morphology of a sample. It makes up the deficiency of an optical microscope and scanning electron microscope, and plays an irreplaceable role in fields such as nanometer materials [8–12], the nano biomedical field [13–16], chemistry [17–20] and microelectronics [21–25]. However, a major constraint on the development of x-ray microscopy is the absence of effective focusing optical elements, while the reflective objective, such as the Schwarzschild objective, provides an effective focusing approach. Extensive research has been done on the reflective objective of the x-ray microscopy, ranging from coating [26–29], design [30–33], to application [16, 34–36].

Research from [37] is a representative example of using the Schwarzschild objective for research on laser plasma. The Schwarzschild objective has a numerical aperture NA of 0.1, and is composed of a spherical surface and an aspherical surface. Under the laser with a wavelength of 13.5 nm, the system can obtain a lateral resolution of 70 nm [37]. When the Schwarzschild objective is used for x-ray research, the mirror surface is usually coated with Mo–Si multilayer film by unidirectional magnetron sputtering. To improve the reflectivity of the mirror, Mo and Si are coated in turn on all the optical elements, and the top layer is always the Mo layer [32, 38].

The reflective objective also plays a very important role in the EUV/UV field. In the EUV/UV band, few varieties of transmissive glass materials are available, and all the available transmissive glass materials have a low transmittance, which limits the use of transmissive objectives. However, the reflective objective is not subject to the limitation of wavelengths, and is free from problems such as chromatic aberration or low transmittance. With the development of UV coating technology [39–43], reflective films used in the EUV/UV band have been relatively mature [44–46]. Therefore, reflective objective working under this spectral band (e.g. the Schwarzschild objective) has been widely used, and has received tremendous attention from fields such as extreme ultra-violet lithography [47–52], micro–nano processing and detection [53–55], and biological environment [53–55].

In 2000, the Lawrence Berkeley National Laboratory, in the United States, developed an extreme ultra-violet lithography micro-exposure tool (MET) that is based on a two-slice reflective objective. The MET has a magnification of 5×, an exposure numerical aperture of 0.3, and an exposure field of view in object space of 1 mm × 3 mm, in which the obscuration radius accounts for 30% [51]. By using an extreme ultra-violet light source with the wavelength of 13.5 nm, the MET successfully etched 25 nm-wide grating lines [51].

In the ultraviolet band, the reflective microscopic objective is used not only in the lithography field, but in the micro–nano processing field [56–59]. Reference [56] presents a representative example of using the Schwarzschild objective for micro–nano processing [56]. Lasers of 193 nm are used to etch a microwell array with a spacing of 500 nm and a diameter of 250 nm. The objective used is a Schwarzschild objective whose numerical aperture is 0.4. Using the Schwarzschild objective in the micro–nano processing field has the advantage that the reflected structure will not absorb the ultraviolet rays, and therefore, the efficiency of light energy utilization is high, and has the disadvantage that the obscuration will have an impact on the final effect of processing.

ENEA Research Laboratories in Italy successfully developed Italy's first EUV lithography device (MET-EGERIA) in 2009. The MET-EGERIA uses a Schwarzschild reflective objective whose numerical aperture is 0.1 as the exposure objective, therefore obtaining a resolution of 90 nm and an exposure field of view of 300 μm × 300 μm [47].

The Schwarzschild reflective objective plays a significant role in the fields of extreme ultra-violet lithography and micro–nano processing, and establishes the foundation for the future exploration of mature extreme ultra-violet lithography technique. The problems of obscuration and field of view shall also be taken into consideration, especially the obscuration, which restricts the further development of the reflective objective. Therefore, four-slice and eight-slice obscuration-free reflective objectives have gradually become the hot spot of research.

In addition to the extreme ultraviolet lithography field, the Schwarzschild reflective objective is widely used in the fields such as wide spectrum imaging and fluorescence research. Reference [55] presents a reflective objective for the ultraviolet excited fluorescence research of bio-aerosol particles [55]. Several laser beams irradiate the bio-aerosol particles simultaneously, and the excited light is collected by the Schwarzschild objective for spectroscopy analysis, during which the advantage of the wide working wavelength of the reflective objective is made full use of.

The Schwarzschild reflective objective has also been used in the near infrared band very well, particularly in the biomedicine field [60–67]. In biological tissues, the near-infrared light has powerful penetrability [68, 69] but causes little damage [70]; therefore, it is a frequently used light source for the observation of *in vivo* tissues. However, there is little optical material that can be used in the infrared wave band, and the reflective microscopy is usually used to conduct the scientific research.

The reflective objective itself has a characteristic of complete achromatism, and can be used in the range from the ultraviolet to the infrared. Therefore, there is a big market need for reflective objectives, and many manufacturers have launched their reflective objectives. Table 1.1 lists the parameters of several common reflective objectives in the market, all of which can be directly used for the fields such as Fourier transform infrared (FTIR) spectrum, a lithography test of semiconductor wafer, lithography, oval film measurement, hyperspectral imaging, thermal imaging microscopy, UVF imaging, etc.

As seen in table 1.1, the working wavelengths of reflective objectives from different manufacturers fall in a wide range from the ultraviolet to the infrared, and the reflective objectives vary somewhat in working wavelengths, which is caused by the difference in reflective films coated on the reflective objectives. The

Table 1.1. Parameters of commercial reflective objectives.

Manufacturer	Model	Numerical aperture	Magnification	Working distance(mm)	Working wavelength (nm)	Obscuration ratio (%)
Thorlabs	LMM-15X-UVV	0.3	15	23.8	200–20 000	25
Thorlabs	LMM-40X-UVV	0.5	40	7.8	200–20 000	22
Edmund	REFLX-15X	0.28	15	24.5	150–11 000	18.9
Edmund	REFLX-25X	0.4	25	14.5	150–11 000	16.7
Newport	50105-01	0.4	15	24	200–20 000	23.6
Newport	50102-01	0.52	36	10.4	200–20 000	21.5

obscuration ratio in the table means the proportion of the region of light obscuring the entire region of the imaging beam. It can be seen that the obscuration ratios of all the reflective objectives are about 20%.

1.4 Summary

In the design model of the aspherical reflective objective in polar coordinate system, the curvature radius of aspheric surface's vertex is coupled with the aspherical coefficient, so the analytical solutions to the radius and the aspherical coefficient cannot be directly obtained. Compared with the spherical reflective objective, the aspherical reflective objective has the advantages of high design freedom, flexible structure and the like, but the increase in the design parameters results in a too complex design model in the polar coordinate system. Therefore, it cannot be used in practical design and analysis. Since Head put forward the design model of an aspherical reflective objective in polar coordinates in 1957, the application of that model has not yet been documented in the open literature. Therefore, obtaining a simple and effective design model of an aspherical reflective objective is one of the important scientific problems in the current design theories of the aspherical reflective objective.

For the coaxial reflection microscopy imaging, the obscuration caused by the secondary mirror of the reflective objective is one of the important factors that affect the image quality. The secondary mirror blocks some parts of the light from the object plane, and only the light at the edge contributes to the imaging. The existence of large obscuration not only reduces the brightness of the image, but also seriously lowers the image quality. The problem of large obscuration caused by the secondary mirror of the reflective objective is one of the major reasons that restrain the development of the reflective objective. The existing reflective objective model can be used to solve the structural parameters of the reflective objective, but the obscuration proportion, which is a key parameter, cannot be effectively controlled. Therefore, establishing a design model of structural parameters of the reflective objective with a constraint on the obscuration ratio is an urgent scientific problem in the design theories of the reflective objective.

References

[1] Smith W J 2005 *Modern Lens Design* (New York: McGraw-Hill) pp 133–57
[2] Schwarzschild K G 1905 Untersuchungen Zur Geometrischen Optik. III *Astronomische Mittheilungen der Königlichen Sternwarte zu Göttingen* (Göttingen: Universität Göttingen) pp 11–5
[3] Artyukov I A 2012 Schwarzschild objective and similar two-mirror systems *Proc. SPIE* **86780A** 1–5
[4] Shealy D L, Wang C and Hoover R B 1995 Optical analysis of an ultra-high resolution two-mirror soft x-ray microscope *J. X-Ray Sci. Technol.* **5** 1–19
[5] Shealy D L, Gabardi D R and Hoover R B *et al* 1989 Design of a normal incidence multilayer imaging x-ray microscope *J. X-Ray Sci. Technol.* **1** 190–206
[6] Head A K 1957 The two-mirror aplanat *Proc. Phys. Soc. Sect.* B **70** 945–51

[7] Burch C 1947 Reflecting microscopes *Proc. Phys. Soc.* **59** 41–7

[8] Vippola M, Valkonen M and Sarlin E *et al* 2016 Insight to nanoparticle size analysis-novel and convenient image analysis method versus conventional techniques *Nanoscale Res. Lett.* **11** 169–76

[9] Trofymchuk I M, Roik N and Belyakova L 2016 Sol-gel synthesis of ordered beta-cyclodextrin-containing silicas *Nanoscale Res. Lett.* **11** 174–9

[10] Siddiqi K S and Husen A 2016 Fabrication of metal nanoparticles from fungi and metal salts: scope and application *Nanoscale Res. Lett.* **11** 98–105

[11] Shpotyuk Y, Ingram A and Shpotyuk O *et al* 2016 Free-volume nanostructurization in Ga-modified As2Se3, glass *Nanoscale Res. Lett.* **11** 1–7

[12] Yang S, Li G and Wang G 2015 Synthesis of Mn_3O_4 nanoparticles/nitrogen-doped graphene hybrid composite for nonenzymatic glucose sensor *Sens. Actuators B Chem.* **221** 172–8

[13] Zhu B, Li Y and Lin Z *et al* 2016 Silver nanoparticles induce HePG-2 cells apoptosis through ROS-mediated signaling pathways *Nanoscale Res. Lett.* **11** 1–8

[14] Weissleder R and Nahrendorf M 2015 Advancing biomedical imaging *Proc. Natl. Acad. Sci. U. S. A.* **112** 14424–8

[15] Nykypanchuk D, Maye M M and Van d L D *et al* 2008 DNA-guided crystallization of colloidal nanoparticles *Nature* **451** 549–52

[16] Sakdinawat A and Attwood D 2010 Nanoscale x-ray imaging *Nat. Photonics* **4** 840–8

[17] Pylypchuk I V, Kołodyńska D and Kozioł M *et al* 2016 Gd-DTPA adsorption on chitosan/magnetite nanocomposites *Nanoscale Res. Lett.* **11** 1–10

[18] Pavlovska O, Vasylechko L and Buryy O 2016 Thermal behaviour of Sm 0.5, R, 0.5 FeO3, (R = Pr, Nd) probed by high-resolution x-ray synchrotron powder diffraction *Nanoscale Res. Lett.* **11** 1–6

[19] Danilchenko V 2016 Ultrafine-grained structure of Fe-Ni-C austenitic alloy formed by phase hardening *Nanoscale Res. Lett.* **11** 1–3

[20] Vesel A, Kovac J and Zaplotnik R 2015 Modification of polytetra fluoroethylene surfaces using H2S plasma treatment *Appl. Surf. Sci.* **357** 1325–32

[21] Savkina R K, Gudymenko A I and Kladko V P *et al* 2016 Silicon substrate strained and structured via cavitation effect for photovoltaic and biomedical application *Nanoscale Res. Lett.* **11** 1–7

[22] Lukianova O A, Krasilnikov V V and Parkhomenko A A *et al* 2016 Microstructure and phase composition of cold isostatically pressed and pressureless sintered silicon nitride *Nanoscale Res. Lett.* **11** 1–6

[23] Boshko O, Dashevskyi M and Mykhaliuk O *et al* 2016 Structure and strength of Iron-Copper-Carbon nanotube nanocomposites *Nanoscale Res. Lett.* **11** 1–8

[24] Hosokai T, Hinderhofer A and Bussolotti F *et al* 2015 Thickness and substrate dependent thin film growth of picene and impact on the electronic structure *J. Phys. Chem.* C **119** 29027–37

[25] Balamurugan C and Lee D W 2015 Perovskite hexagonal $YMnO_3$ nanopowder as P-type semiconductor gas sensor for H_2S detection *Sens. Actuators B Chem.* **221** 857–66

[26] Zhu J, Tu Y and Yuan Y *et al* 2014 Thermal stability of Co/C multilayers *Mater. Res. Express* **1** 046503–9

[27] Tu Y, Zhu J and Li H *et al* 2014 Structural changes induced by thermal annealing in Cr/C multilayers *Appl. Surf. Sci.* **313** 341–5

[28] Artyukov I A, Feschenko R M and Vinogradov A V *et al* 2010 Soft x-ray imaging of thick carbon-based materials using the normal incidence multilayer optics *Micron* **41** 722–8

[29] Barbee T W JR 1986 Multilayers for x-ray optics *Opt. Eng.* **25** 893–915

[30] Artyukov I, Bugayev Y and Devizenko O *et al* 2009 x-ray Schwarzschild objective for the carbon window (λ~4.5 nm) *Opt. Lett.* **34** 2930

[31] Artioukov I A and Krymski K M 2000 Schwarzschild objective for soft x-rays *Opt. Eng.* **39** 2163–70

[32] Horikawa Y, Mochimaru S and Iketaki Y 1992 Design and fabrication of the Schwarzschild objective for soft x-ray microscopes *Proc. SPIE-The Int. Soc. Opt. Eng.* **1720** 217–25

[33] Shealy D L, Jiang W and Hoover R B 1991 Design and analysis of aspherical multilayer imaging x-ray microscope *Opt. Eng.* **30** 1094–9

[34] Lai M, Gao Y and Yuan B *et al* 2015 Remarkable superelasticity of sintered Ti-Nb alloys by Ms adjustment via oxygen regulation *Mater. Des.* **87** 466–72

[35] Malyutin A A 1997 Analysis of the applications of the Schwarzschild objective in the soft x-ray and VUV spectral ranges. 1. Conditions for fifth-order aplanatism *Quantum Electron.* **27** 90–5

[36] Artioukov I A, Vinogradov A V and Asadchikov V E *et al* 1995 Schwarzschild soft-x-ray microscope for imaging of nonradiating objects *Opt. Lett.* **20** 2451–3

[37] Nechai A N, Pestov A E and Polkovnikov V N *et al* 2016 x-ray optical system for imaging laser plumes with a spatial resolution of up to 70 nm *Quantum Electron.* **46** 347–52

[38] Changxin Z 1997 Reflective x-ray microscope using laser plasma as a light source *OME Inf.* **6** 15–7

[39] Chkhalo N I and Salashchenko N N 2013 Next generation nanolithography based on Ru/Be and Rh/Sr multilayer optics *AIP Adv.* **3** 279–87

[40] Barysheva M M, Pestov A E and Salashchenko N N *et al* 2012 Precision imaging multilayer optics for soft x-rays and extreme ultraviolet bands *Physics-Uspekhi* **55** 727–47

[41] Wang H C 2004 Analysis of the reflective performance of EUV multilayer under the influence of capping layer *Acta Phys. Sin.* **53** 2368–72

[42] Bajt S, Alameda J B and Barbee T W *et al* 2002 Improved reflectance and stability of Mo-Si multilayers *Opt. Eng.* **41** 1797–804

[43] Bajt S, Alameda J B and Barbee T W *et al* 2001 Improved reflectance and stability of Mo/Si multilayers *Proc. SPIE* **41** 65–75

[44] Dawei L, Chun G and Yundong Z 2012 Preparation of aluminium mirror coatings of the vacuum ultraviolet *Acta Opt. Sin.* **02** 327–31

[45] Shuyi G 2008 Study on the characteristics and related technology of vacuum ultraviolet reflective film *PhD Thesis* University of Science and Technology of China pp 8–12.

[46] Shuyi G, Xiangdong X and Yilin H 2006 Review on highly reflecting mirrors for vacuum ultraviolet and x-ray *Chin J. Vac. Sci. Technol.* **06** 459–68

[47] Flora F, Bollanti S and Lazzaro P D 2009 The first Italian micro exposure tool for EUV lithography: design guidelines and experimental result 115–9

[48] Lazzaro P D 2009 Excimer-laser-driven EUV plasma source for single-shot projection lithography *IEEE Trans. Plasma Sci.* **37** 475–80

[49] Bollanti S, Di Lazzaro P and Flora F *et al* 2008 A laser-plasma clean soft x-ray source for projection microlithography *XVII International Symposium on Gas Flow and Chemical Lasers and High Power Lasers* (Bellingham, WA: International Society for Optics and Photonics) pp 713116–9

[50] Bollanti S, Di Lazzaro P and Flora F *et al* 2006 Conventional and modified Schwarzschild objective for EUV lithography: design relations *Appl. Phys. B-Lasers Opt.* **85** 603–10

[51] Naulleau P, Goldberg K A and Cain J P *et al* 2005 EUV microexposures at the ALS using the 0.3-NA MET projection optics *Emerging Lithographic Technologies IX* **5751** pp 56–63

[52] Taylor J, Sweeney D and Hudyma R *et al* 2000 EUV microexposure tool (MET) for near-term development using a high NA projection system *Proceedings of the 2nd International EUVL Workshop (October)* pp 115–9

[53] Juschkin L, Maryasov A and Herbert S *et al* 2011 EUV dark-field microscopy for defect inspection *The 10th International Conference on x-ray Microscopy* vol 1365 (Melville, NY: AIP Publishing) pp 265–8

[54] Barkusky F, Bayer A and Peth C *et al* 2009 Direct photoetching of polymers using radiation of high energy eensity from a table-top extreme ultraviolet plasma source *J. Appl. Phys.* **105** 14906–7

[55] Pan Y L, Holler S and Chang R K *et al* 1999 Single-shot fluorescence spectra of individual micrometer-sized bioaerosols illuminated by a 351-or a 266-nm ultraviolet laser *Opt. Lett.* **24** 116–8

[56] Karstens R, Gödecke A and Prießner A *et al* 2016 Fabrication of 250-nm-hole arrays in glass and fused silica by UV laser ablation *Opt. Laser Technol.* **83** 16–20

[57] Wang Z, Wang X and Mu B *et al* 2012 Nanoscale patterns made by using a 13.5-nm Schwarzschild objective and a laser produced plasma source *SPIE Photonics Europe* (Bellingham, WA: International Society for Optics and Photonics) pp 843012-1–12.

[58] Wang X, Mu B and Jiang L *et al* 2011 Fabrication of nanoscale patterns in lithium fluoride crystal using a 13.5 nm Schwarzschild objective and a laser produced plasma source *Rev. Sci. Instrum.* **82** 123702

[59] Barkusky F, Bayer A and Peth C *et al* 2007 Compact EUV source and Schwarzschild objective for modification and ablation of various materials *International Congress on Optics and Optoelectronics* (Bellingham, WA: International Society for Optics and Photonics) pp 65860B-1–12.

[60] Pan Y L, Hill S C and Santarpia J L *et al* 2014 Spectrally-resolved fluorescence cross sections of aerosolized biological live agents and simulants using five excitation wavelengths in a BSL-3 laboratory *Opt. Express* **22** 8165–89

[61] Kaucikas M, Barber J and Van Thor J J 2013 Polarization sensitive ultrafast mid-IR pump probe micro-spectrometer with diffraction limited spatial resolution *Opt. Express* **21** 8357–70

[62] Birarda G, Ravasio A and Suryana M *et al* 2016 IR-Live: fabrication of a low-cost plastic microfluidic device for infrared spectromicroscopy of living cells *Lab Chip* **16** 1644–51

[63] Mattson E C, Unger M and Clède S *et al* 2013 Toward optimal spatial and spectral quality in widefield infrared spectromicroscopy of IR labelled single cells *Analyst* **138** 5610–8

[64] Kaucikas M, Barber J and Van Thor J J 2013 Polarization sensitive ultrafast mid-IR pump probe micro-spectrometer with diffraction limited spatial resolution *Opt. Express* **21** 8357–70

[65] Artyukov I A 2012 Schwarzschild objective and similar two-mirror systems *Short-Wavelength Imaging and Spectroscopy Sources* pp 112–9

[66] Ekgasit S, Pattayakorn N and Tongsakul D *et al* 2007 A novel ATR FT-IR micro-spectroscopy technique for surface contamination analysis without interference of the substrate *Anal. Sci.* **23** 863–8

[67] Burghoff J, Will M and Nolte S *et al* 2005 Photonics in silicon using mid-IR femtosecond pulses *Lasers and Applications in Science and Engineering* (Bellingham, WA: International Society for Optics and Photonics) pp 245–52

[68] Mao W, Chunyan L and Yunfei S 2013 Research of near-infrared small living animal fluorescence imaging system *Acta Opt. Sin.* **6** 239–44

[69] Zhang Y, Hong G and Zhang Y *et al* 2012 Ag$_2$S quantum dot: a bright and biocompatible fluorescent nanoprobe in the second near-infrared window *ACS Nano* **6** 3695–702

[70] Xumeng W 2014 Highly stable NIR fluorescent sensors and their application in bioimaging *PhD thesis* East China University of Science and Technology, Shanghai pp 4–12.

IOP Publishing

Elliptical Mirrors
Applications in microscopy
Jian Liu

Chapter 2

Apodization factor and linearly polarized light focusing properties of elliptical mirror

Mengzhou Li, Jian Liu, Min Ai, Cien Zhong and Jiubin Tan

2.1 Introduction

The traditional scalar diffraction theory is a good approximation for the research on the propagation of light in most cases; however, when the energy distribution of diffraction field is closely related to the polarized state of light, for example the focusing of light waves in an elliptical mirror system with a high aperture angle, the vector diffraction theory must be applied to achieve good accuracy.

The vector diffraction theory of an aspherical lens was first proposed by E Wolf and B Richard [1, 2]. Subsequently, Richard Barakat, C J R Sheppard, A Bovin, Min Gu, A S van de Nes, Rakeshkesh Kumar Singh *et al* published papers on research based on the vector diffraction theory [3–11]. In 1919, V S Ignatowsky modeled the parabolic mirror system with the vector diffraction theory [12] in his paper, and it was almost the first paper introducing the vector diffraction theory of a reflective system. The fundamental bases for E Wolf and B Richard to build a vector diffraction model of aplanatic lenses are that, when the light is at the interface,

(1) the included angle between the electric or magnetic vectors and the light in a meridian plane remains unchanged;
(2) the electric or magnetic vectors are always on the same side of the meridian plane.

Then the three-dimensional expression of a focusing electromagnetic field was deduced according to the energy conservation principle and based on the Debye–Wolf integral.

In the vector theory, an apodization factor depicting wave front amplitude changes is used to represent the energy conversion relationship when light waves pass through a lens or mirror, and it is also one of the keys to developing the focusing theories of elliptical mirrors with a large aperture angle. This chapter deduces the apodization factor of an elliptical mirror according to the energy conservation principle, provides various expressions of apodization factor function with respect to different variable parameters, and finally deduces a three-dimensional electromagnetic field analytical expression of the focus region of an elliptical mirror system under the linearly polarized light incident conditions based on the Debye–Wolf vector diffraction theory. The principle of developing the apodization factor of elliptical mirrors in this chapter was inspired by the work of V S Ignatowsky on parabolic mirrors, and his work helped us a lot in the initial understanding of the physical meaning behind the apodization factor.

2.2 Elliptical mirror model

In this chapter, a complete ellipsoid refers to the closed surface formed by rotating an ellipse around its major axis for a complete cycle. A mirror whose surface is in the shape of a portion of the ellipsoid is regarded as an elliptical mirror. This book mainly studies the focusing and imaging characteristics of the elliptical mirror, and especially, the elliptical mirror in this book specially refers to the mirror with rotational symmetry. As shown in figure 2.1, the ellipsoid in a rectangular coordinate system can be described as:

$$\frac{z^2}{a^2} + \frac{x^2 + y^2}{b^2} = 1, \tag{2.1}$$

where, a and b are respectively the major and minor semi-axes of the ellipsoid.

The distance between two conjugate focal points is:

$$|F_1F_2| = 2c = 2\sqrt{a^2 - b^2}. \tag{2.2}$$

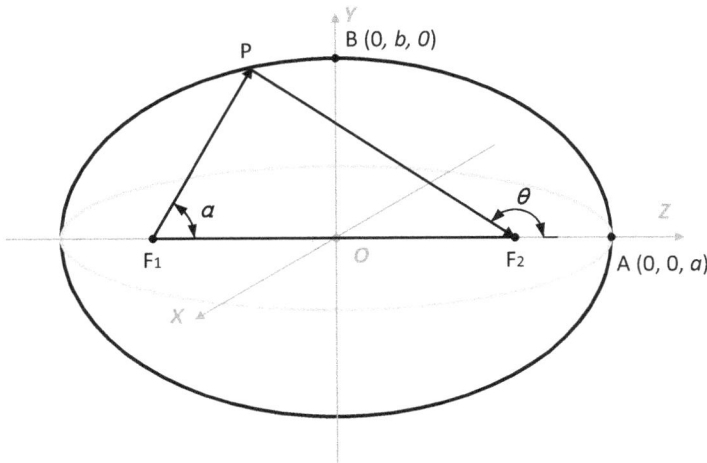

Figure 2.1. Parameters in an elliptical mirror.

According to the definition of quadric surface eccentricity, the elliptical eccentricity e can be expressed as:

$$e = \frac{c}{a} = \frac{\sqrt{a^2 - b^2}}{a}, \quad (0 < e < 1). \tag{2.3}$$

In addition, another two variables are involved in the practical fabrication of elliptical surfaces, i.e. radius of curvature R and aspherical coefficient K, which are respectively defined as follows:

$$R = \frac{b^2}{a} \lim_{x \to \infty}, \tag{2.4}$$

$$K = \frac{b^2}{a^2} - 1 = -e^2. \tag{2.5}$$

Based on the geometrical optics theory, light from a focal point of an ellipsoid will be focused at the other focal point after being reflected by the ellipsoid. This process is a perfect imaging process, but the field of view is only an ideal point. In order to research the angular relationship between the incident ray and the reflected ray, without loss of generality, it is assumed that both the incident ray and the reflected ray are on the YOZ plane, and P is the reflecting point, as shown in figure 2.1. The included angle between the incident ray and the positive direction of Z axis is called the incidence angle α, and the included angle between the reflected ray and the negative direction of Z axis is called the illumination angle θ.

According to the Law of Sines, there exists

$$\frac{|PF_1|}{\sin \theta} = \frac{|PF_2|}{\sin \alpha} = \frac{|F_1F_2|}{\sin (\theta - \alpha)}. \tag{2.6}$$

To eliminate $|PF_1|$ and $|PF_2|$ with the property of ellipse:

$$|PF_1| + |PF_2| = 2a, \tag{2.7}$$

thereby it can be obtained that

$$\frac{2a}{\sin \theta + \sin \alpha} = \frac{2c}{\sin (\theta - \alpha)}. \tag{2.8}$$

Formula (2.8) shows the angular relationship between the incident ray and the reflected ray that passes through the focal points.

2.3 Apodization factor

In the analysis of a lens with a high numerical aperture, as shown in figure 2.2, the thin lens focuses the plane waves into spherical waves, and the amplitude distribution of the wavefront of plane waves that are incident on the thin lens is $p(r)$, in which r represents the height of the incident ray to the optic axis. The wavefront that passes through the lens is a focusing spherical wave whose amplitude distribution is $p(\theta)$, and θ denotes the included angle between the exit ray and the optic axis. Then the mapping relationship between $p(r)$ and $p(\theta)$ is referred as the apodization factor of the lens. Commonly, in the case of a low numerical aperture lens, we have $p(\theta) \approx p(r)$; however, when the numerical aperture is higher than 0.75, $p(r)$ and $p(\theta)$ differ greatly, and therefore, an apodization factor is needed to correct the amplitude distribution. The apodization factor reflects the change of light field amplitude before and after the incident light from different locations passing through the optical component.

Under the sine condition, the apodization factor of an aplanatic lens deduced by Richards and Wolf in 1959 is:

$$p_{lens}(\theta) = \sqrt{\cos\theta}. \tag{2.9}$$

The apodization factor of a parabolic mirror is similar to that of the aplanatic lens, as shown in figure 2.3, and was first proposed in a paper by the Soviet physicist V S Ignatovsky in 1919.

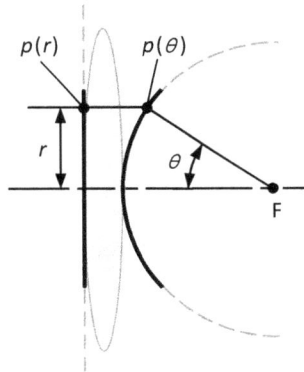

Figure 2.2. Apodization factor of lens.

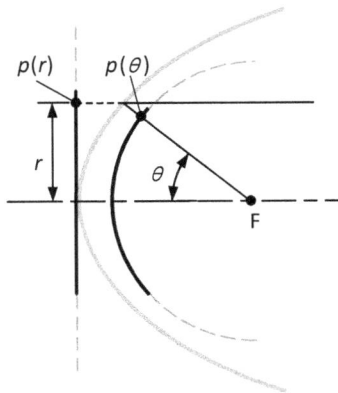

Figure 2.3. Apodization factor of the parabolic mirror.

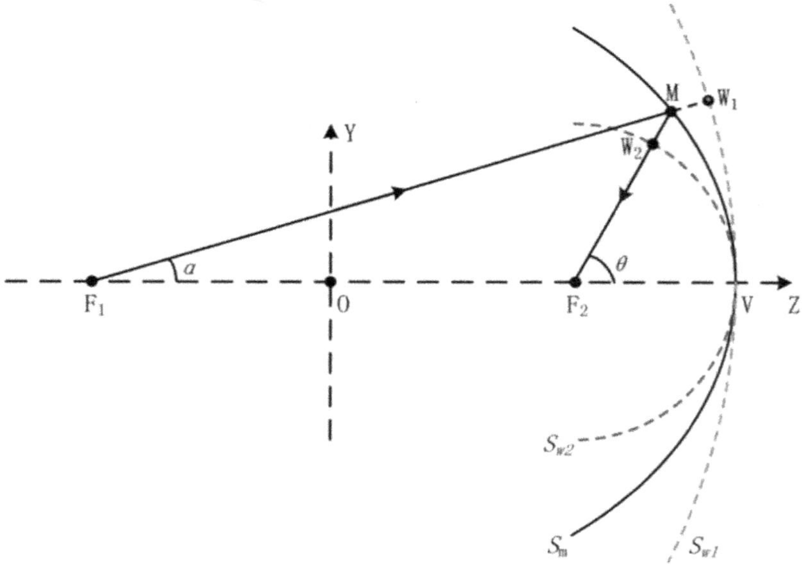

Figure 2.4. Principle of solving apodization factor of elliptical mirror [13].

$$p_{parabolic}(\theta) = \frac{2}{1 + \cos\theta}. \qquad (2.10)$$

Incident waves of the lens and the parabolic mirror are both plane waves, but for the elliptical mirror, because of the dual focal points, the incident wave is a spherical wave, and the exit wave is still a spherical wave. Therefore, compared with the lens and the parabolic mirror, the apodization factor of the elliptical mirror is deduced in a slightly different way [13]. The principle of solving the apodization factor of an elliptical mirror is illustrated in figure 2.4. After being reflected by the elliptical mirror, a spherical wave S_{w1} emitted from the focal point F_1 is changed into another spherical wave S_{w2}, which will focus at the other focal point F_2. Thus, the apodization factor $p(\theta)$ is defined as the ratio of the amplitude of reflected wave to that of the incident wave at any point M on the elliptical mirror. For an ideal mirror, we ignore the energy loss caused by absorption and reflection. Then we take a very small annular region that is adjacent to the point M, and according to the energy conservation principle,

$$A_1^2 \delta S_{W1} = A_2^2 \delta S_{W2}, \qquad (2.11)$$

we can obtain the apodization factor

$$p(\theta) = A_2/A_1 = \sqrt{\delta S_{w1}/\delta S_{w2}}, \qquad (2.12)$$

where δS_{w1} and δS_{w2} are respectively the area elements of the wavefront before and after the reflection.

According to the geometrical relationship, we can obtain

$$\delta S_{w1} = 2\pi f_1 \sin \alpha d\alpha, \tag{2.13}$$

$$\delta S_{w2} = 2\pi f_2 \sin \theta d\theta, \tag{2.14}$$

where $f_1 = a + c$ and $f_2 = a - c$ are respectively the radii of focal spherical surfaces S_{w1} and S_{w2}. Therefore, the amplitude apodization factor of elliptical mirror can be expressed as:

$$p(\theta) = \frac{a + c}{a - c} \sqrt{\frac{\sin \alpha d\alpha}{\sin \theta d\theta}}. \tag{2.15}$$

When the focusing angle θ_{max} is no larger than $\pi/2$, the angle α can be expressed as the function of the angle θ as

$$\left.\begin{aligned} z &= (ca^2tg^2\theta + ab^2\sqrt{1 + tg^2\theta})/(a^2tg^2\theta + b^2), \\ \alpha &= ctg[(z - c)tg\theta/z]. \end{aligned}\right\} \tag{2.16}$$

When the focusing angle θ_{max} is greater than $\pi/2$, the functional relationship between the angle α and the angle θ is changed into

$$\left.\begin{aligned} z &= (ca^2tg^2\theta - ab^2\sqrt{1 + tg^2\theta})/(a^2tg^2\theta + b^2), \\ \alpha &= ctg[(z - c)tg\theta/z]. \end{aligned}\right\} \tag{2.17}$$

The apodization factor reflects the change in energy distribution caused by reflection or transmission at the exit pupil. If the incident angle is small, the value of the apodization factor $p(\theta)$ will be close to 1, which indicates that from wave S_{w1} to wave S_{w2}, the energy density remains unchanged. This kind of energy change can be vividly interpreted as the change of light beam projection region. However, when the apodization factor $p(\theta)$ is greater or less than 1, in the very small region, the energy after reflection or transmission will be compressed or stretched. The energy density will also increase or decrease correspondingly. Figure 2.5 shows the curves of apodization factors of the lens, parabolic mirror and elliptical mirror as a function of angle θ, where the structure parameters of the elliptical mirror are $a = 500$ mm and $c = 300$ mm. As can be seen from the figure, when the angle θ approaches $\pi/2$ from 0, the apodization factors of the mirrors gradually increase to 2 from 1, and the apodization factor of the lens gradually decreases from 1 to 0. Different variation trends of the apodization factor indicate that in the energy density, the reflective system performs better in delivering high-frequency components than the lens system, as high incident angle wave components represent high-frequency components in the space frequency domain.

For the elliptical mirror, the apodization factor is also related to elliptical parameters a, b and c, not just the function of focusing angle θ. This is where the elliptical mirror differs from the lens and the parabolic mirror. Figure 2.6 shows the

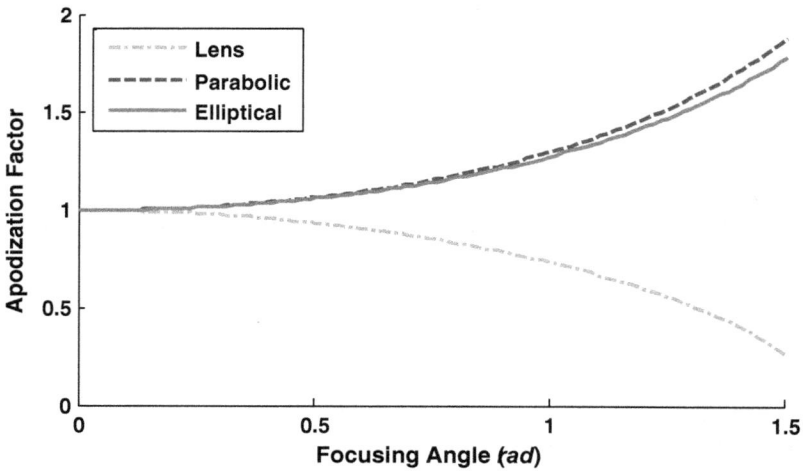

Figure 2.5. Apodization factors of lens, parabolic mirror and elliptical mirror ($e = 0.6$).

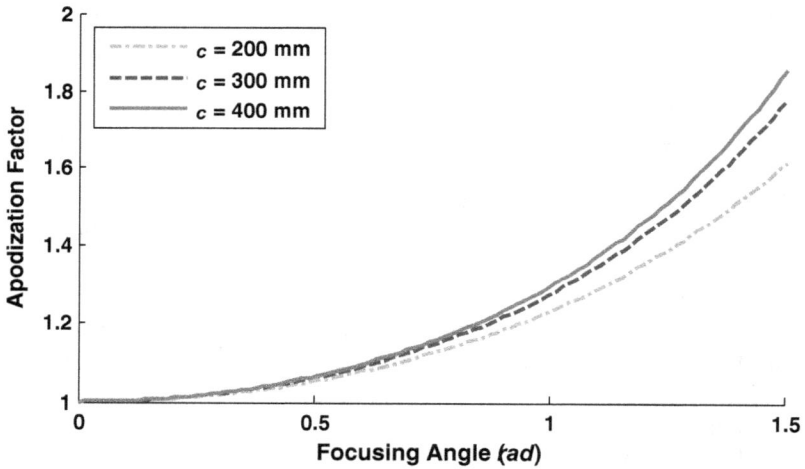

Figure 2.6. Apodization factor as a function of centrifugal distance c of elliptical mirror ($a = 500$ mm).

change curves of apodization factor as a function of angle θ at different centrifugal distances c. When c gradually decreases, two focal points of the ellipsoid move toward each other, and the elliptical mirror gets closer to the shape of a spherical mirror. On the other hand, the apodization factor decreases and approaches to 1, which means that the energy change is gradually reduced and in the extreme case, there is no amplitude change between the incident wave and the exit wave.

2.4 Apodization factor under different parametric variables

To be noticed, the expression of the apodization factor of the elliptical mirror given in formula (2.15) is not an analytical expression but a function of several variables, which is disadvantageous to further calculation or numerical simulation. In order to simplify this formula, the nature properties of the elliptical mirror can be taken full advantage of to change it into a univariate analytical function.

2.4.1 Apodization factor in terms of z

For any point M on the elliptical mirror, the ratio of the distance between point M and focal point F_2 to the distance between point M and the plane $z = a/e$, where the corresponding directrix locates, is a definite value e. Similarly, the ratio of the distance between point M and focal point F_1 to the distance between point M and the plane $z = -a/e$ is also equal to e, as shown in figure 2.7.

Using the geometrical relationship of the triangle $F_1 M F_2$, we can obtain the formula as follows:

$$\alpha = \arccos\left(\frac{z_M + ae}{a + ez_M}\right), \theta = \arccos\left(\frac{z_M - ae}{a - ez_M}\right), \tag{2.18}$$

where z_M is the Z position of point M in the coordinate system of figure 2.7. According to formula (2.18), we can further obtain the formulas as follows:

$$\frac{d\alpha}{dz} = -\frac{1}{\sqrt{1 - (z + ae)^2/(a + ez)^2}} \cdot \frac{a - ae^2}{(a + ez)^2},$$

$$\frac{d\theta}{dz} = -\frac{1}{\sqrt{1 - (z - ae)^2/(a - ez)^2}} \cdot \frac{a - ae^2}{(a - ez)^2}. \tag{2.19}$$

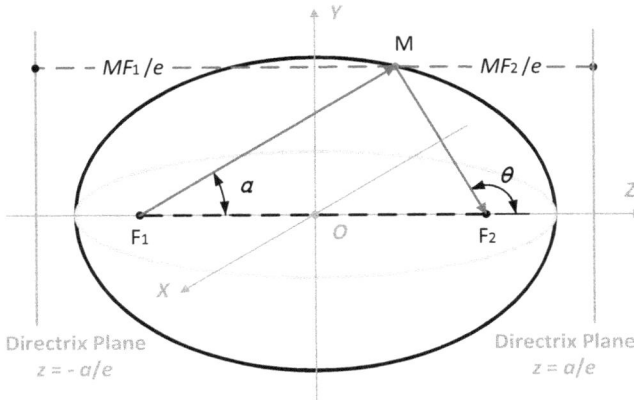

Figure 2.7. Geometrical relationships in the ellipsoid.

Then, according to the Law of Sines, formula (2.15) can be simplified as:

$$p(\theta) = \frac{a + ae}{a - ae}\sqrt{\frac{\sin\alpha d\alpha}{\sin\theta d\theta}} = \frac{1 + e}{1 - e}\sqrt{\frac{(a - ez)d\alpha}{(a + ez)d\theta}} = \frac{1 + e}{1 - e}\frac{(a - ez)}{(a + ez)}, \tag{2.20}$$

i.e.

$$p(z) = \frac{1 + e}{1 - e}\frac{(a - ez)}{(a + ez)}. \tag{2.21}$$

Formula (2.21) is an analytical expression, and is continuous within the aperture angle interval of $[0, \pi]$ [14]. Compared to formula (2.15), it is applicable for the analysis of the focusing properties of an elliptical mirror with an aperture angle greater than $\pi/2$ with which focusing process includes both forward and backward illumination, while formula (2.15) is not. Moreover, simulation based on formula (2.15) will also be time consuming, because the function is not analytic. But with formula (2.21), the simulation will be much faster.

2.4.2 Apodization factor in terms of θ

Since the integral variable for the simulation of focusing properties of elliptical mirror is usually the angular variable, simplifying the apodization factor as an expression using an incidence angle or emergence angle as an independent variable will provide greater simplicity and convenience for the simulation or calculation [15]. Thus it is very natural for us to develop the apodization factor in terms of θ or α.

According to formula (2.8), we have

$$a\sin(\theta - \alpha) = c(\sin\alpha + \sin\theta). \tag{2.22}$$

Applying the trigonometric sum-to-product identities, we can obtain

$$a\sin\frac{\theta - \alpha}{2} = c\sin\frac{\theta + \alpha}{2}. \tag{2.23}$$

With some calculation, it can be obtained that

$$a\left(\tan\frac{\theta}{2} - \tan\frac{\alpha}{2}\right) = c\left(\tan\frac{\theta}{2} + \tan\frac{\alpha}{2}\right), \tag{2.24}$$

$$(a - c)\tan\frac{\theta}{2} = (a + c)\tan\frac{\alpha}{2}. \tag{2.25}$$

After differentiating both sides of formula (2.25), we can obtain:

$$\frac{a - c}{2}\frac{1}{\cos^2(\theta/2)}d\theta = \frac{a + c}{2}\frac{1}{\cos^2(\alpha/2)}d\alpha, \tag{2.26}$$

$$\frac{d\theta}{d\alpha} = \frac{a+c}{a-c} \cdot \frac{\cos^2(\theta/2)}{\cos^2(\alpha/2)}. \tag{2.27}$$

According to the trigonometric relationship, there is

$$\frac{\sin\theta}{\sin\alpha} = \frac{\tan(\theta/2)}{\tan(\alpha/2)} \cdot \frac{\tan^2(\alpha/2)+1}{\tan^2(\theta/2)+1}. \tag{2.28}$$

By synthesizing formulas (2.12)–(2.14), (2.28) and (2.27), finally we have

$$l^2(\alpha) = \frac{\tan^2(\theta/2)+1}{\tan^2(\alpha/2)+1} \cdot \frac{\cos^2(\alpha/2)}{\cos^2(\theta/2)}. \tag{2.29}$$

By using the trigonometric relationship of $\cos^2\theta = 1/(\tan^2\theta + 1)$ and the double angle formula, formula (2.29) can be transformed as follows

$$l^2(\alpha) = \frac{\cos^4(\alpha/2)}{\cos^4(\theta/2)} = \left[\frac{\tan^2(\theta/2)+1}{\tan^2(\alpha/2)+1}\right]^2 = \left(\frac{1+\cos\alpha}{1+\cos\theta}\right)^2. \tag{2.30}$$

Therefore, with formula (2.25), the apodization factor of the elliptical mirror can be written as

$$l(\alpha) = \frac{a^2 + c^2 - 2ac\cos\alpha}{(a-c)^2} \quad or \quad l(\theta) = \frac{(a+c)^2}{a^2 + c^2 + 2ac\cos\theta}. \tag{2.31}$$

Formula (2.30) and (2.31) provide two expression forms of the apodization factor, one of which is an implicit expression which is simple but relevant to both the incidence angle α and the focusing angle θ, and shall be used in combination with formula (2.22) or (2.25); and the other of which is an explicit expressio,n which is relevant to either the incidence angle α or the focusing angle θ. These forms of apodization factor will bring great convenience to both theoretical analysis and numerical simulation on focusing properties of elliptical mirror. They are also applicable to the situation where the elliptical mirror has an aperture angle greater than $\pi/2$.

2.5 Focusing properties based on the vector theory

E Wolf pointed out in a paper that if an optical system meets the following two conditions, (1) the size of exit pupil is much greater than the wavelength of incident light, and (2) the distance between the imaging space and the exit pupil is much greater than the wavelength of incident light, then the electromagnetic field of the imaging space can be expressed by formulas (2.32) and (2.33) respectively:

$$e(x, y, z) = -\frac{ik}{2\pi} \iint_\Omega \frac{a(s_x, s_y)}{s_z} e^{ik[\phi(s_x, s_y) + s_x x + s_y y + s_z z]} ds_x ds_y, \tag{2.32}$$

$$h(x, y, z) = -\frac{ik}{2\pi} \iint_\Omega \frac{b(s_x, s_y)}{s_z} e^{ik[\phi(s_x, s_y)+s_x x+s_y y+s_z z]} ds_x ds_y, \qquad (2.33)$$

where a and b are respectively the electric field strength factor and the magnetic field strength factor, and k is the wave number. Since the magnetic field can be obtained from the electric field in the same location, typically only the electric field of the focus region needs to be obtained.

2.5.1 Vector theories

The elliptical mirror system has a unique characteristic of bifocus conjugation. When one focal point is used as the light source, the other one can be used as the point detection end. To obtain a point source at the focal point of ellipsoid, place an aberration-free lens meeting the sine condition in front of the ellipsoid, and make sure the back focal point of the lens coincides with the front focal point of the ellipsoid. Then the incident plane waves are focused by the lens as the point light source, reflected by the ellipsoid, and ultimately focused at the other focal point. Figure 2.8 shows the geometrical schematic diagram of the elliptical mirror focusing system. The point lighting is obtained by the foregoing aberration-free lens and focused at the focal point F_1. To calculate the electric field near the focal point F_2, we define a series of vectors. The focal point F_1 is taken as the coordinate origin, and the plane where the light ray and the optical axis Z locate is taken as the meridian plane. s_L, s_0 and s_1 are unit vectors along the propagation direction of the light ray in the meridian plane. g_L, g_0 and g_1 are the corresponding vertical unit vectors, and α, β, θ and γ are the corresponding polar angles between the vectors and the Z axis. The maximum values of α and θ are determined by the numerical apertures of lens and elliptical mirror. M is any point on the elliptical mirror. φ is the included angle between the meridian plane F_1MF_2 and the X axis.

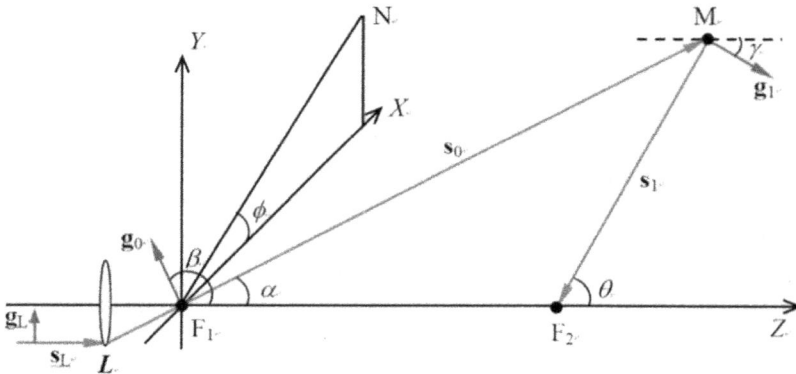

Figure 2.8. Vector relationship of focusing by the elliptical mirror [13].

Vectors g_L, g_0, g_1, s_L, s_0 and s_1 can be respectively expressed as:

$$g_L = \cos \phi \mathbf{i} + \sin \phi \mathbf{j} , \qquad (2.34)$$

$$g_0 = \cos \alpha \cos \phi \mathbf{i} + \cos \alpha \sin \phi \mathbf{j} - \sin \alpha \mathbf{k}, \qquad (2.35)$$

$$g_1 = -\cos \theta \cos \phi \mathbf{i} - \cos \theta \sin \phi \mathbf{j} + \sin \theta \mathbf{k}, \qquad (2.36)$$

$$s_L = \mathbf{k}, \qquad (2.37)$$

$$s_0 = \sin \alpha \cos \phi \mathbf{i} + \sin \alpha \sin \phi \mathbf{j} + \cos \alpha \mathbf{k}, \qquad (2.38)$$

$$s_1 = -\sin \theta \cos \phi \mathbf{i} - \sin \theta \sin \phi \mathbf{j} - \cos \theta \mathbf{k}. \qquad (2.39)$$

Light waves at the interface of the lens meet the following conditions: (1) the included angle between the electric vector and the principal ray remains unchanged; (2) the electric vectors are always on the same side of the meridian plane. However, when the light waves arrive at the interface of the elliptical mirror, the components of the electric vectors remain unchanged in the parallel meridian plane but point to an opposite direction in the vertical meridian plane due to phase jump. We assume that light waves polarize along the x direction in the incident ray polarization plane, which is expressed as $e_L^0 = l_0(\theta)(1, 0, 0)$. When the light waves pass through the lens, the electric field at the exit pupil of lens is expressed as e_L^1, the electric field at the sphere centered at the front focal point of the elliptical mirror is expressed as e_F^0, and the electric field at the sphere centered at the back focal point of the elliptical mirror is expressed as e_F^1.

For the relationship at the lens, the electric vectors are always perpendicular to the direction of light propagation during the light wave propagation. When the light beams incident on the lens surface are linearly polarized plane waves, the electric field e_L^1 at the exit pupil of lens can be expressed as:

$$e_L^1 = \sqrt{\cos \alpha}\left[\beta g_0 + \gamma(g_0 \times s_0)\right], \qquad (2.40)$$

where, $\sqrt{\cos \alpha}$ is the apodization factor of the lens that meets the sine condition, and coefficients β and γ can be obtained by the principle met at the interface:

$$\left.\begin{array}{l} \beta = e_L^1 \cdot g_0 = e_L^0 \cdot g_L \\ \gamma = e_L^1 \cdot (g_0 \times s_0) = e_L^0 \cdot (g_L \times s_L) \end{array}\right\}. \qquad (2.41)$$

Light waves at the exit pupil of lens arrive at the elliptical mirror after the free propagation for a distance, and only the amplitude and phase of the electric field change, while its direction remains unchanged. The electric field at the sphere centered at the front focal point of the elliptical mirror is expressed as follows:

$$e_F^0 = \frac{f_L}{a+c} e_L^1 \times e^{j2k(f_L+a+c)}. \tag{2.42}$$

We assume that, after the reflection by the ellipsoid, the apodization factor of the ellipsoid is $p(\theta)$, and the electric field at the sphere centered at the back focal point of the elliptical mirror is expressed as follows:

$$e_F^1 = p(\theta)\big[\eta g_1 + \mu(g_1 \times s_1)\big]. \tag{2.43}$$

Coefficients η and μ can be obtained by the principle met when the light waves pass through the reflective plane:

$$\left.\begin{aligned}
\eta &= e_F^1 \cdot g_1 = e_F^0 \cdot g_0 = e_L^1 \cdot g_0 = e_L^0 \cdot g_L \\
\mu &= e_F^1 \cdot (g_1 \times s_1) = -e_F^0 \cdot (g_0 \times s_0) = -e_L^1 \cdot (g_0 \times s_0) \\
&= -e_L^0 \cdot (g_L \times s_L)
\end{aligned}\right\}. \tag{2.44}$$

Coefficients η and μ are respectively the amplitude of radial components and azimuth components of the electric field at the back focal point of the elliptical mirror.

2.5.2 Three-dimensional expression of the focused electric field

For the elliptical mirror, according to formulas (2.32) and (2.33), the electric field adjacent to focal point F_2 in the imaging space can be expressed as:

$$e_s = -\frac{ik}{2\pi}\int_0^{2\pi}\int_0^{\theta_{max}} \sqrt{\cos\alpha}\, \frac{f_L}{a+c} e^{ik(f_L+a+c)} e_F^1 e^{iks_1\cdot r} \sin\theta d\theta d\phi, \tag{2.45}$$

where, k is the wave number, θ_{max} is the maximum focusing angle of elliptical mirror, and r is the location vector of the observation point, i.e.,

$$r = \rho_s \sin\theta_s \cos\phi_s i + \rho_s \sin\theta_s \sin\phi j + \rho_s \cos\theta_s k. \tag{2.46}$$

When the focusing angle θ_{max} is equal to or less than π, according to formulas (2.34)–(2.44) and (2.31) and by using the following equation

$$\left.\begin{aligned}
\int_0^{2\pi} \sin(n\phi)e^{ik\rho\cos(\phi-\phi_s)}d\phi &= 2\pi i^n J_n(\rho)\sin(n\phi_s) \\
\int_0^{2\pi} \cos(n\phi)e^{ik\rho\cos(\phi-\phi_s)}d\phi &= 2\pi i^n J_n(\rho)\cos(n\phi_s)
\end{aligned}\right\}. \tag{2.47}$$

Formula (2.45) can be simplified as

$$
\left.\begin{aligned}
e_s^x &= iA\big(I_0 + I_2 \cos\left(2\phi_s\right)\big), \\
e_s^y &= iAI_2 \sin\left(2\phi_s\right), \\
e_s^z &= -2AI_1 \cos\phi_s.
\end{aligned}\right\}, \tag{2.48}
$$

where, the constant term is $A = \dfrac{\pi f_L}{\lambda(a-c)}e^{ik(f_L+a+c)}$, and the integral term is as shown in the following formula (2.49)

$$
\left.\begin{aligned}
I_0 &= \int_0^{\theta_{\max}} p(\theta)\sin\theta(1+\cos\theta)J_0(k\rho_s \sin\theta_s \sin\theta)e^{-ik\rho_s \cos\theta_s \cos\theta}d\theta \\
I_1 &= \int_0^{\theta_{\max}} p(\theta)\sin^2\theta J_1(k\rho_s \sin\theta_s \sin\theta)e^{-ik\rho_s \cos\theta_s \cos\theta}d\theta \\
I_2 &= \int_0^{\theta_{\max}} p(\theta)\sin\theta(1-\cos\theta)J_2(k\rho_s \sin\theta_s \sin\theta)e^{-ik\rho_s \cos\theta_s \cos\theta}d\theta
\end{aligned}\right\}. \tag{2.49}
$$

When given the maximum aperture angle of the elliptical mirror, formula (2.48) shows the three-dimensional expression of the focused electric field of elliptical mirror at the maximum aperture angle $\theta_{\max} \in (0,\pi)$ generated with x linearly polarized light.

2.5.3 Numerical simulation of the focusing field

2.5.3.1 Intensity distribution
To facilitate the numerical simulation analysis, the cylindrical coordinates $r/\lambda = r_s \cos\phi_s \boldsymbol{i} + r_s \sin\theta_s \boldsymbol{j} + z_s \boldsymbol{k}$ are used, and formula (2.49) is transformed into formula (2.50):

$$
\left.\begin{aligned}
I_0 &= \int_0^{\theta_{\max}} p(\theta)\sin\theta(1+\cos\theta)J_0(2\pi r_s \sin\theta)e^{-i2\pi z_s \cos\theta}d\theta \\
I_1 &= \int_0^{\theta_{\max}} p(\theta)\sin^2\theta J_1(2\pi r_s \sin\theta)e^{-i2\pi z_s \cos\theta}d\theta \\
I_2 &= \int_0^{\theta_{\max}} p(\theta)\sin\theta(1-\cos\theta)J_2(2\pi r_s \sin\theta)e^{-i2\pi z_s \cos\theta}d\theta
\end{aligned}\right\}. \tag{2.50}
$$

After the linearly polarized light along the x direction is focused by the elliptical mirror, the electric field energy density (related to the intensity of the electric field) adjacent to the focal point F_2 is defined as w_e, which can be expressed as follows by the electric field of the focus region:

$$
w_e = \frac{1}{16\pi}(\boldsymbol{e}_s \cdot \boldsymbol{e}_s^*). \tag{2.51}
$$

The electric field $\boldsymbol{e}_s$ is given by formulas (2.48) and (2.50).

a) Intensity Change along x Axis

b) Intensity Change along y Axis

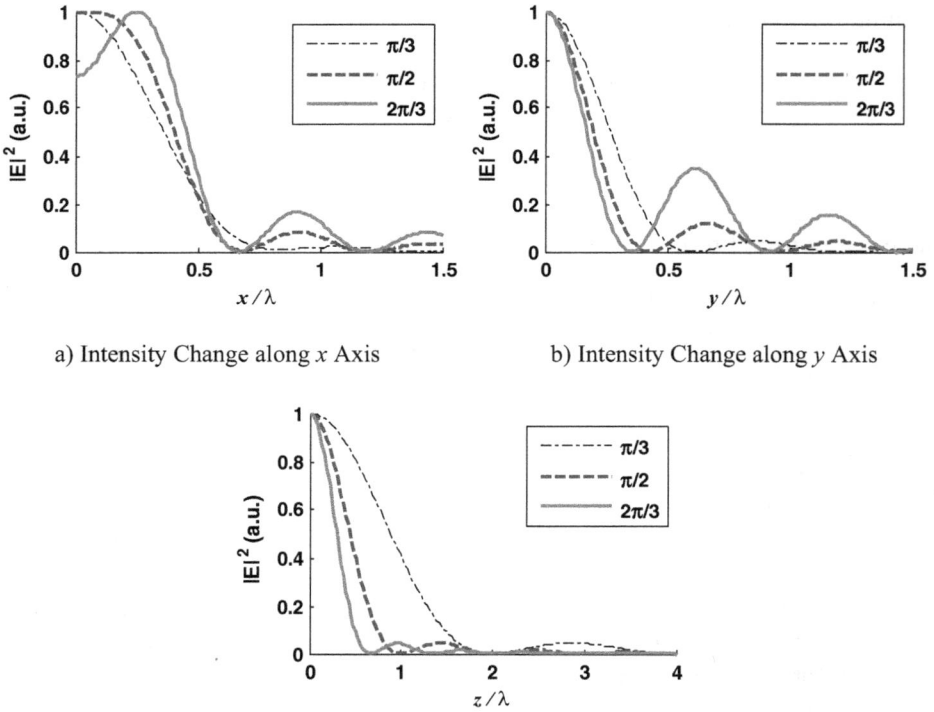

Figure 2.9. Changes of electric field intensity in focus region along the axes.

We assume the elliptical mirror has a major axis $a = 500$ mm and a minor axis $b = 400$ mm, and the focusing angles θ_{max} are respectively $\pi/3$, $\pi/2$ and $2\pi/3$. Then figure 2.9 shows the intensity curves of electric field of the focus region on three coordinate axes, among which figure 2.9(a) shows the change curves on x axis, figure 2.9(b) shows the change curves on y axis, and figure 2.9(c) shows the change curves on z axis.

As shown in figure 2.9, when the focusing angle of the elliptical mirror increases, the electric field intensity of the focus region along three coordinate axes decreases in the width of the main lobe, among which the width of the main lobe on the z axis decreases most. The results show that as the aperture angle of elliptical mirror increases, both the lateral and the axial resolutions increase, and of the two, the axial resolution increases more. It is worth mentioning that the amplitude of side lopes also increases along with the decrease in the width of the main lobe, especially along the y axis. In addition, as shown in figure 2.9(a), along the x axis the width of the main lobe slightly gets narrowed but the slope becomes much sharper and the width of half width of half intensity actually gets wider. Especially, there appears a pit near the origin when the aperture angle becomes larger than $\pi/2$, and this may be due to the polarization of the incident light which causes the asymmetry of the focus spot. The asymmetry effect will become more apparent if the aperture angle gets very large.

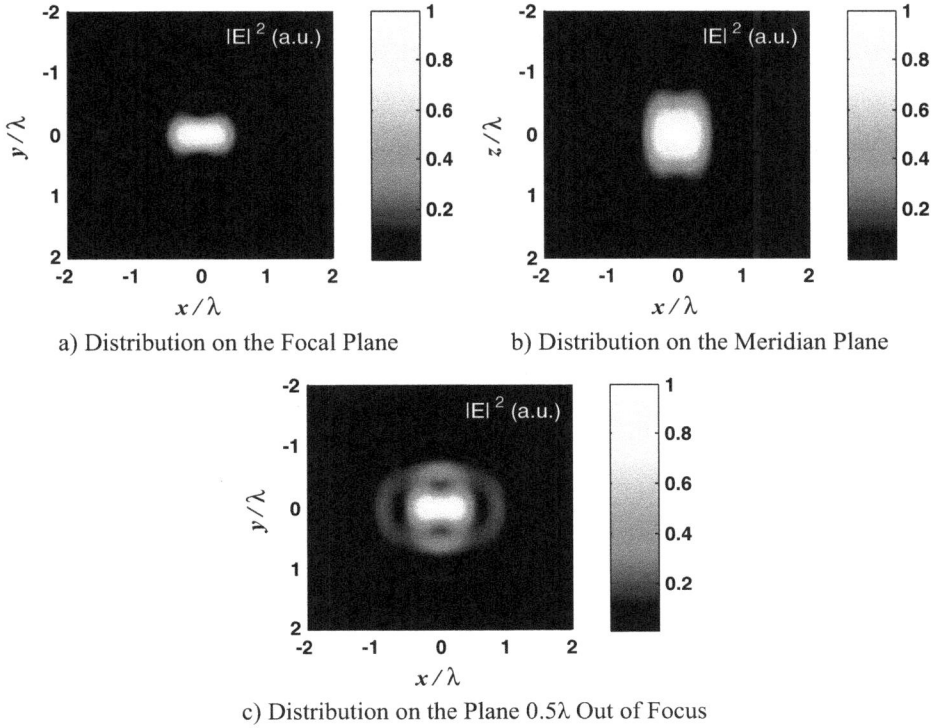

a) Distribution on the Focal Plane

b) Distribution on the Meridian Plane

c) Distribution on the Plane 0.5λ Out of Focus

Figure 2.10. Distribution of normalized electric field intensity under x polarized illumination in the focus region of the elliptical mirror.

If the aperture angle of the elliptical mirror θ_{max} is assumed to be $\pi/2$, and the major and minor axis are 500 m and 400 mm respectively, then, on the focal plane, the intensity distribution of focused electric field generated with x polarized light in the elliptical mirror is as shown in figure 2.10, among which figure 2.10(a) shows the intensity distribution of focused electric field on the focal plane xoy, figure 2.10(b) shows the intensity distribution of focused electric field on the meridian plane xoz, and figure 2.10(c) shows the intensity distribution of focused electric field on the plane 0.5 λ out of focus.

According to figure 2.10(a), the intensity distribution of focused electric field is unsymmetrical on the x and y axes, and the spot size on the y axis is smaller, indicating that the resolution on the y axis is higher. This is mainly caused by the incident light's linearly polarization along x axis, and we can obtain a symmetric and much narrower spot on the focal plane when utilizing a radially polarized illumination, which will be introduced in chapter 3; as shown in figure 2.10(b), the intensity distribution of focused electric field is also unsymmetrical in the lateral and axial directions, and the axial size of the spot size is larger than the lateral size, which is similar to the focusing of an ordinary lens; as is seen from figure 2.10(c), the intensity distribution of focused electric field on the plane 0.5 λ out of focus is similar to that on the focal plane, except that the relative amplitude decreases.

2.5.3.2 Polarization state

The polarization state of the electric field around the focus region can be obtained based on its three-dimensional analytical expression. Any complex electric vector can be expressed in the following form composed of a real part and an imaginary part,

$$e_s = p + iq, \tag{2.52}$$

where, p and q are respectively the real part and the imaginary part of the electric vector, and are a pair of conjugate semi-diameters of the polarization ellipse of the electric vector. Similarly, integral terms I_0, I_1 and I_2 can also be expressed in the similar form, i.e.

$$I_n = I_n^r + iI_n^i \quad (n = 0, 1, 2). \tag{2.53}$$

Dividing each complex number into a real part and an imaginary part can facilitate the calculation, and according to formulas (2.48) and (2.50), we can obtain:

$$\left.\begin{aligned}
p_x &= -A(I_0^i + I_2^i \cos 2\phi_s), \\
p_y &= -AI_2^i \sin 2\phi_s, \\
p_z &= -2AI_1^r \cos \phi_s.
\end{aligned}\right\}, \tag{2.54}$$

$$\left.\begin{aligned}
q_x &= A(I_0^r + I_2^r \cos 2\phi_s), \\
q_y &= AI_2^r \sin 2\phi_s, \\
q_z &= -2AI_1^i \cos \phi_s.
\end{aligned}\right\}. \tag{2.55}$$

For an imaging point on the focal plane, $z_s = 0$. According to formula (2.20), integral terms I_0, I_1 and I_2 are real numbers. Therefore, formulas (2.54) and (2.55) can be simplified as

$$\left.\begin{aligned}
p_x &= 0, \\
p_y &= 0, \\
p_z &= -2AI_1 \cos \phi_s.
\end{aligned}\right\}, \tag{2.56}$$

$$\left.\begin{aligned}
q_x &= A(I_0 + I_2 \cos 2\phi_s), \\
q_y &= AI_2 \sin 2\phi_s, \\
q_z &= 0.
\end{aligned}\right\}. \tag{2.57}$$

According to formulas (2.56) and (2.57), the point on the focal plane meets $p \cdot q = 0$, which means the conjugate semi-diameters are at right angles to each other. Thus, on the focal plane, p and q are the semi-axes of the polarization ellipse of the electric vector. Furthermore, the p axis is perpendicular to the focal plane, while the q axis is in the focal plane. Therefore, for any point on the focal plane, the plane of the polarization ellipse of the electric vector is always perpendicular to the focal plane. The included angle χ between the plane of the polarization ellipse and the xoz plane is given by the formula as follows:

$$\tan \chi = \frac{q_y}{q_x} = \frac{I_2 \sin 2\phi_s}{I_0 + I_2 \cos 2\phi_s}. \tag{2.58}$$

The ratio ψ of two axes of the polarization ellipse is given by the formula as follows:

$$\psi = \frac{|p_z|}{\sqrt{q_x^2 + q_y^2}} = \frac{2\,|I_1 \cos \phi_s|}{\sqrt{I_0^2 + I_2^2 + 2I_0 I_2 \cos 2\phi_s}}. \tag{2.59}$$

According to formulas (2.58) and (2.59), we can draw the distributions of the angle χ and the ratio ψ of the electric vector on the focal plane. For the elliptical mirror whose aperture angle $\theta_{\max}$ is equal to $\pi/2$, major axis a is 500 mm, and minor axis b is 400 mm, figure 2.11 shows the distribution of the polarization state of focused electric field on the focal plane generated with x linearly polarized light, among which figure 2.11(a) shows the distribution of the included angle χ, between the plane of the polarization ellipse and the xoz plane, of the electric field on the focal plane, and figure 2.11(b) shows the distribution of the ratio ψ of two axes of the polarization ellipse of the electric field on the focal plane.

According to figure 2.11(a), all the values on the y axis are equal to zero, indicating that the electric field at any point on the y axis polarizes in the xoz plane. Values at the focal point and various points in small regions adjacent to the focal points are all equal to zero, indicating that the electric fields at various points within small central regions polarize in the xoz plane, too. For each point on the z axis,

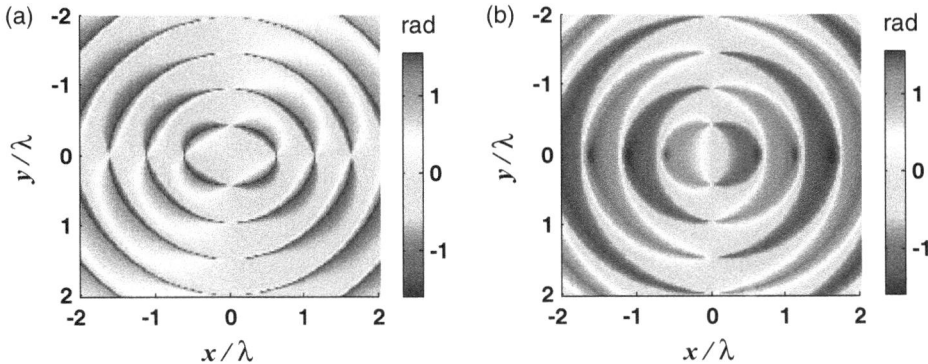

Figure 2.11. Distribution of polarization state of focused electric field on the focal plane. (a) Included angle between the plane of the polarization ellipse and the xoz plane; (b) ratio of two axes of polarization ellipse ($\arctan(\psi)$).

$r_s = 0$, and therefore, $I_1 = 0$ and $I_2 = 0$. The e_s^y and e_s^z components are all equal to zero. Among all the electric vectors, only e_s^x components are not equal to zero, that is for each point on the optical axis, the electric vector is linearly polarized along the x direction. According to figure 2.11(b), all the values at various points on the y axis are equal to zero, indicating that the semi-axis, perpendicular to the focal plane, of the polarization ellipse is zero, and together with the result we got from figure 2.11(a), we know the polarizations of the electric filed at these points are along the x-direction. Similarly, values in regions adjacent to the focal points are also small, indicating that the semi-axis parallel with the focal plane is the major axis of the polarization ellipse, and the length of it is much greater than that of the other semi-axis.

2.5.3.3 Influence of surface parameter on intensity distribution

According to the foregoing analysis, when the surface parameters of the elliptical mirror change, the apodization factor changes correspondingly. Suppose the aperture angle θ_{max} of the elliptical mirror is equal to $\pi/2$, and the major axis a is still 500 mm. When the centrifugal distance c is 400 mm, 300 mm and 200 mm, the corresponding intensity distribution of the focused electric field is shown in figure 2.12, among which the figure 2.12(a) shows the change curves on x axis, figure

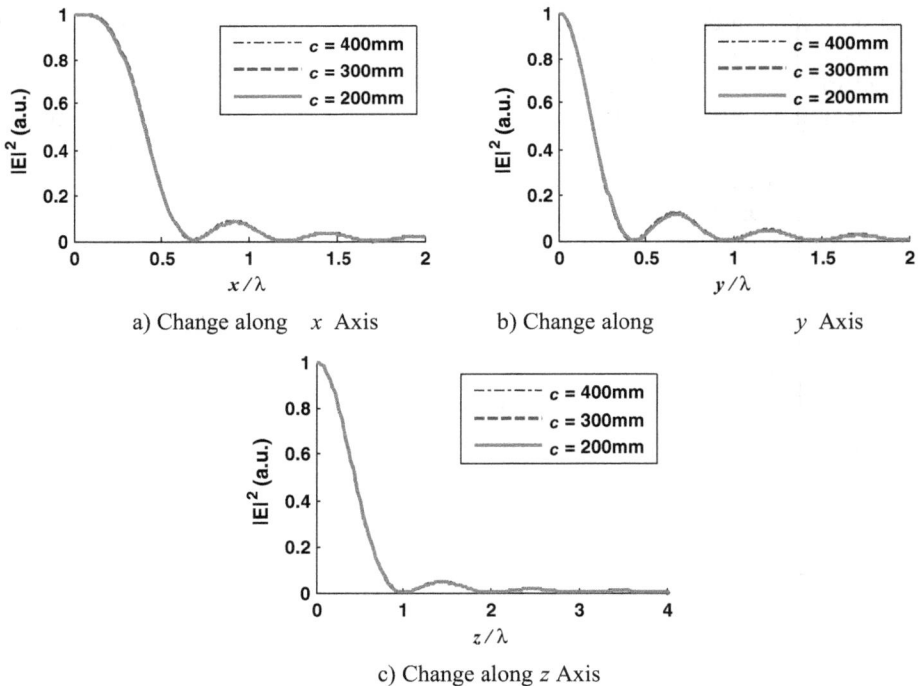

a) Change along x Axis

b) Change along y Axis

c) Change along z Axis

Figure 2.12. Intensity distribution of focused electric field as a function of surface parameter of the elliptical mirror (a = 500 mm).

2.12(b) shows the change curves on y axis, and figure 2.12(c) shows the change curves on z axis.

As shown in figure 2.12, when the centrifugal distance of the elliptical mirror changes, the intensity distribution curves of focused electric field generally coincide on three rectangular coordinate axes, indicating that when the aperture angle and the major axis of the elliptical mirror are given, the change in the centrifugal distance has little influence on the intensity distribution of the focused electric field.

2.6 Comparison of focusing properties among elliptical mirror, parabolic mirror and lens

Detailed analysis of the intensity distribution of the focused electric field generated with x linearly polarized light in the elliptical mirror is made in the previous section. What is the difference in the intensity distribution of focused electric field in elliptical mirror, parabolic mirror and aplanatic lens? We assume that the numerical aperture of each of these three systems is 0.95, i.e. the maximum focusing angle is arcsin (0.95). Figure 2.13 shows the intensity distribution of focused electric field generated with linearly x polarized light in the elliptical mirror, parabolic mirror and aplanatic lens. Since the curves on the positive and negative semi-axes are symmetrical, only

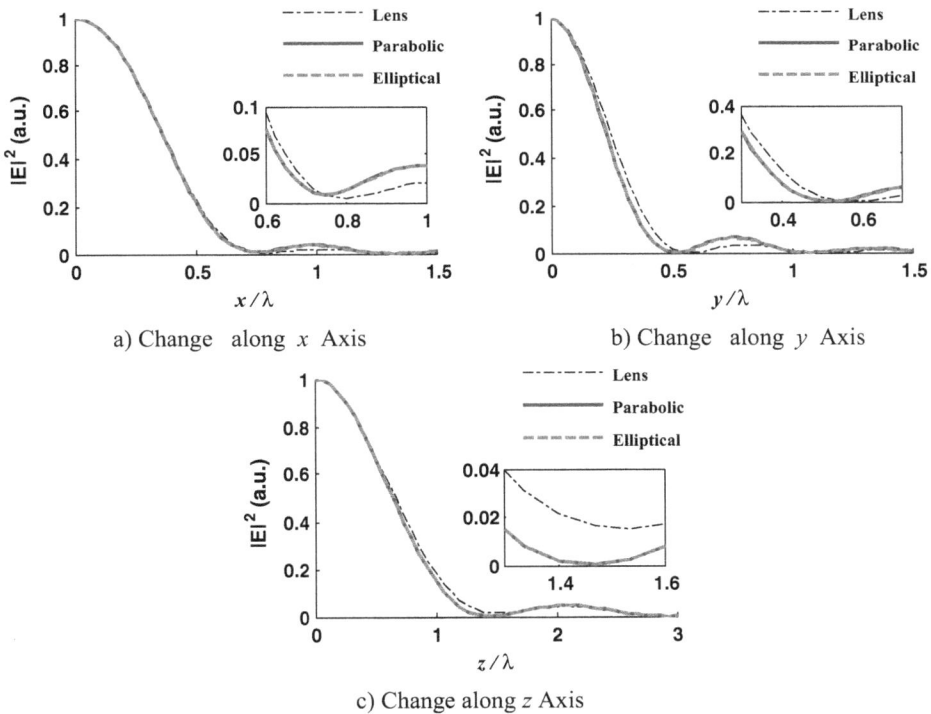

a) Change along x Axis

b) Change along y Axis

c) Change along z Axis

Figure 2.13. Intensity distribution curves of focused electric field generated by linearly polarized light in three systems.

Table 2.1. Comparison of main lobe widths of focused electric field intensity distribution curves in three systems.

	Elliptical mirror h_e	Parabolic mirror h_p	Aplanatic lens h_l	h_m/h_l
x Axis	0.75	0.75	0.8	6.25%
y Axis	0.52	0.51	0.58	12.07%
z Axis	1.45	1.45	1.52	4.61%

the envelope curve on the positive axis is provided. Figure 2.13(a) shows the change curves on x axis, figure 2.13(b) shows the change curves on y axis, and figure 2.13(c) shows the change curves on z axis.

According to figure 2.13, with x direction polarized light, elliptical mirror and parabolic mirror generate a narrower main lobe width than lens, indicating that under the same numerical aperture angle, the second order aspherical mirror has smaller lateral and axial focal spots than the aplanatic lens.

Figure 2.13 qualitatively shows that with linearly polarized light, the second-order aspherical mirror can provide higher lateral and axial resolution in light probe scanning microscopy. Then to quantitatively discuss the difference in focal spots of the linearly polarized light in three systems, table 2.1 compares the main lobe widths of focused electric field intensity distribution curves of elliptical mirror, parabolic mirror and aplanatic lens under linearly polarized (along x axis) illumination. In the table, *he*, *hp* and *hl* respectively represent the main lobe widths of the elliptical mirror, parabolic mirror and aplanatic lens. h_m in the last column represents the main lobe width of the second order aspherical mirror (elliptical mirror and parabolic mirror), and h_l represents the main lobe width of the aplanatic lens. It is known from table 2.1 that compared with the aplanatic lens, the second order aspherical mirror reduces the focal spot size by 12.07% in y direction, and reduces the focal spot size by 6.25% and 4.61% respectively along x axis and z axis.

2.7 Summary

This chapter mainly introduces the apodization factor of elliptical mirrors, its various forms in terms of different parameters, and its application in the research of focusing properties of the elliptical mirror. Firstly, we developed the basic concept of the apodization factor of the elliptical mirror based on the energy conservation principle. Then to adapt different calculations and simulation situations, we deduced different expressions for the apodization factor, and tried to simplify them and make them analytic and explicit. Finally, we used the apodization factor to calculate the focused electric fields in the focus region of the elliptical mirror under the linearly polarized illumination, and analyzed them from aspects like intensity distribution, surface parameters influences, comparisons with parabolic mirror and lens, etc. As we can see through this chapter, the differences between the focusing properties of the elliptical mirror and the parabolic mirror or even the lens are mainly caused by their different apodization factors, while their integral equations for calculating the focused electric fields are almost the same except for the apodization functions. Thus it can inspire us to find links between the focusing properties and the apodization functions, and then we can specially design the apodization function, by choosing proper parameters of the elliptical mirror, to meet our specific needs about focusing properties. Moreover, it also reminds us that we can introduce wavefront shaping or pupil filtering-like techniques to change the apodization function and create new focusing properties, and this is not limited in elliptical mirrors.

References

[1] Wolf E 1959 Electromagnetic diffraction in optical systems. I. An integral representation of the image field *Proc. R. Soc.* A **253** 349–57

[2] Richards B and Wolf E 1959 Electromagnetic diffraction in optical systems. II. Structure of the image field in an aplanatic system *Proc. R. Soc.* A **253** 358–79

[3] Barakat R 1987 Diffracted electromagnetic fields in the neighborhood of the focus of a paraboloidal mirror having a central obscuration *Appl. Opt.* **26** 3790–5

[4] Sheppard C J R 1978 Electromagnetic field in the focal region of wide-angular annular lens and mirror systems *Opt. Acoustics* **2** 163–6

[5] Bovin A and Wolf E 1965 Electromagnetic field in the neighborhood of the focus of a coherent beam *Phys. Rev.* **138** 1561–5

[6] Sheppard C J R, Choudhury A and Gannaway J 1977 Electromagnetic field near the focus of wide-angular lens and mirror systems *Opt. Acoustics* **1** 129–32

[7] Bovin A, Dow J and Wolf E 1967 Energy flow in the neighborhood of the focus of a coherent beam *J. Opt. Soc. Am.* **57** 1171–5

[8] Sheppard C J R and Gu M 1993 Imaging by a high aperture optical system *J. Mod. Opt.* **40** 1631–51

[9] Sheppard C J R and Matthews H J 1987 Imaging in high-aperture optical systems *J. Opt. Soc. Am.* **4** 1354–60

[10] Nes A S, Billy L, Pereira S F and Braat J M 2004 Calculation of the vectorial field distribution in a stratified focal region of a high numerical aperture imaging system *Opt. Express* **12** 1281–93

[11] Singh R K, Senthilkumaran P and Singh K 2008 Tight focusing of optical beams *Invertis J. Sci. Technol.* **1** 197–230

[12] Ignatovsky V S 1920 Diffraction by a parabolic mirror having arbitrary opening *Trans. Opt. Inst. Petrograd* **1** 5 (in Russian)

[13] Liu J, Tan J and Wilson T *et al* 2012 Rigorous theory on elliptical mirror focusing for point scanning microscopy *Opt. Express* **20** 6175–84

[14] Liu J, Ai M and Zhang H *et al* 2013 Focusing properties of elliptical mirror with an aperture angle greater than π *Opt. Eng.* **53** 061606

IOP Publishing

Elliptical Mirrors
Applications in microscopy
Jian Liu

Chapter 3

Focusing characteristic of polarized light

He Zhang, Jian Liu, Min Ai and Jiubin Tan

Imaging with an extra high aperture angle ($>\pi$) is a situation unique to reflective imaging, and does not exist in the lens imaging. The discussion of this kind of issue is mainly to explore new theoretic approaches of high-frequency imaging. An elliptical mirror can enable aperture angle to reach π or even go beyond π, which means tighter focal spots may be obtained. Compared with objective lenses, elliptical mirrors can be easily used for the extension of aperture angles. What is more, using elliptical mirrors with aperture angles greater than π, the specimen can be put in forward and backward illuminations.

Besides enlarging the NA of a focusing optical element, using polarized beams is another way to shape the focusing spot. A variety of beams with spatially inhomogeneous polarization, such as radial polarization, azimuthal polarization and generalized cylindrical polarization, are defined as cylindrical-vector (CV) beams, because they are all axially symmetric beams to the full vector electromagnetic wave equation. Due to this inherent axial symmetry, the focusing properties of CV beams differ from those of linearly and circularly polarized beams. Most essentially, for radially polarized beams, a tight focusing spot under diffraction limit can be obtained and it consists of a strong longitudinally polarized field as well. Moreover, for azimuthally polarized beams, former knowledge indicates that a doughnut-shaped focal light field can be acquired. With these interesting properties, focusing of the CV beam, especially the focusing of a radially polarized beam in high aperture, becomes attractive in studies such as particle trapping, high-resolution imaging and laser cutting that require either a strong longitudinal filed or a tight focusing.

This chapter firstly researches an elliptical mirror with an extra-high aperture angle to provide a geometry expression and an apodization factor expression of the elliptical mirror, then takes the Debye–Wolf vector diffraction integral as the basic model to research the vector focus imaging characteristic of the elliptical mirror with an extra-high aperture angle under different polarized light incident conditions (including circularly polarized light and cylindrically polarized light), and analyzes the change in system focusing characteristics as the geometric parameters of the elliptical mirror change. The theoretical model established is not only applicable to the analysis of the elliptical mirror with an aperture angle larger than π, but also to the analysis of the existing elliptical mirror with an aperture angle smaller than π. Therefore, the theoretical model established can be universally used.

3.1 Basic model of an elliptical mirror

The existing apodization factor expression of an elliptical mirror is not suitable for analyzing the focusing characteristic when the aperture angle exceeds π, and the existing ellipsoid geometric equation cannot intuitively reflect the geometric characteristics of the ellipsoid; therefore, to address these problems, this chapter determines suitable forms of geometric equation expression for an elliptical mirror, and deduces an apodization factor expression for the elliptical mirror with an aperture angle larger than π.

As shown in figure 3.1, the elliptical mirror in a rectangular coordinate system can be expressed as follows [1, 2]:

$$\frac{x^2 + y^2}{b^2} + \frac{z^2}{a^2} = 1 \tag{3.1}$$

where, a and b respectively represent the major axis and minor axis of the ellipsoid, and the spacing between the two conjugate focal points of the ellipsoid can be expressed as:

$$|F_1F_2| = 2c = 2\sqrt{a^2 - b^2}. \tag{3.2}$$

On the other hand, according to the geometric definition of eccentricity of a hyperbola, the elliptical eccentricity e can be expressed as:

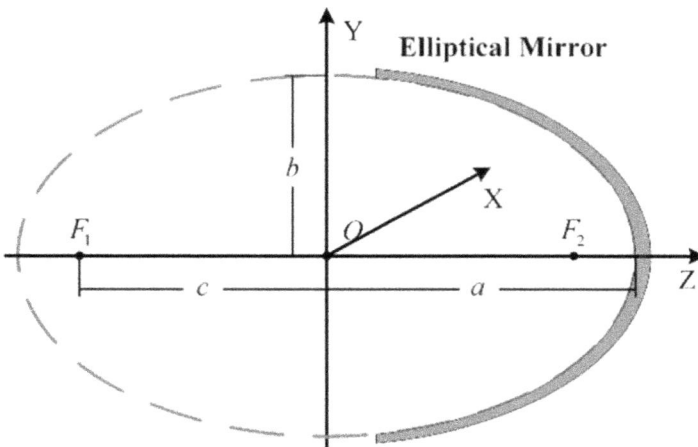

Figure 3.1. Geometric schematic diagram of the elliptical mirror.

$$e = \frac{c}{a} = \frac{\sqrt{a^2 - b^2}}{a}, \qquad (0 < e < 1). \tag{3.3}$$

Then the ellipsoid equation can be rewritten as

$$\frac{x^2 + y^2}{a^2 - e^2 a^2} + \frac{z^2}{a^2} = 1. \tag{3.4}$$

According to the theoretical model in chapter 2, if formula (3.1) is used as the ellipsoid equation, then the apodization factor of the elliptical mirror can be specifically expressed as [3–5]:

$$p(\theta) = \frac{a + c}{a - c} \sqrt{\frac{\sin \alpha \, d\alpha}{\sin \theta \, d\theta}}, \tag{3.5}$$

where, α is the emergence angle of incident spherical wavefront at the focal point F_1, and satisfies:

$$\alpha = \arctan[(z - c)\tan \theta / z], \tag{3.6}$$

and [1]

$$\begin{cases} z = (ca^2 \tan^2 \theta + ab^2 \sqrt{1 + tg^2 \theta})/(a^2 \tan^2 \theta + b^2), & (\theta < \pi/2) \\ z = (ca^2 \tan^2 \theta - ab^2 \sqrt{1 + tg^2 \theta})/(a^2 \tan^2 \theta + b^2). & (\theta \geqslant \pi/2) \end{cases} \tag{3.7}$$

It is evident that the apodization factor expression of the elliptical mirror given by formula (3.5) is not an analytical expression, moreover, repeatedly calling formulas (3.6) and (3.7) during the numerical calculation severely reduces the efficiency of numerical simulation. More importantly, the apodization factor expression of formula (3.5) cannot effectively analyze the focusing characteristic of elliptical mirror with an aperture angle greater than π. Therefore, it is necessary to give the analytical expression of apodization factor of the elliptical mirror, in order to analyze the focusing characteristic of the elliptical mirror system with an aperture angle greater than π.

Besides the references to the existing theoretical model, it is found, during the analysis of the geometrical relationship of the ellipsoid, that, as shown in figure 3.2, for any point M on the elliptical mirror, the ratio of the distance between M and the focal point F_2 to the distance between M and the plane $z = a/e$ where the corresponding directrix locates is a constant value e. In the same way, the ratio of the distance between M and the focal point F_1 to the distance between M and the plane $z = -a/e$ where the corresponding directrix locates is also the constant value e.

Using the geometrical relationship of the triangle $F_1 M F_2$, we can obtain the formula as follows:

$$\alpha = \arccos\left(\frac{z_M + ae}{a + ez_M}\right), \quad \theta = \arccos\left(\frac{z_M - ae}{a - ez_M}\right), \tag{3.8}$$

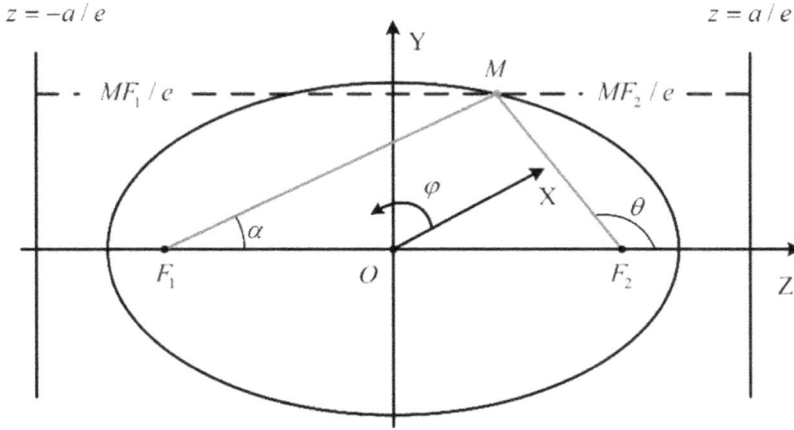

Figure 3.2. Schematic diagram of geometrical relationship of ellipsoid. Adapted from [4].

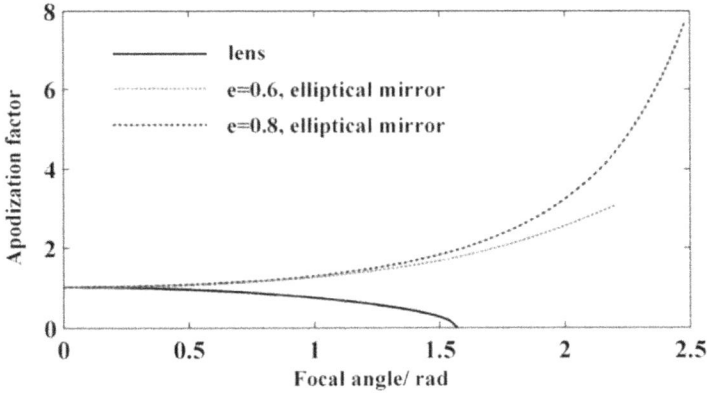

Figure 3.3. Change curves of apodization factors of lens and elliptical mirrors with different eccentricities.

where z_M is the Z position of point M in the coordinate system of figure 3.3. According to formula (3.8), we can further obtain the formulas as follows:

$$\frac{d\alpha}{dz} = -\frac{1}{\sqrt{1 - (z + ae)^2/(a + ez)^2}} \cdot \frac{a - ae^2}{(a + ez)^2},$$

$$\frac{d\theta}{dz} = -\frac{1}{\sqrt{1 - (z - ae)^2/(a - ez)^2}} \cdot \frac{a - ae^2}{(a - ez)^2}. \tag{3.9}$$

According to the Law of Sines, formula (3.5) can be simplified as:

$$p(\theta) = \frac{a + ae}{a - ae} \sqrt{\frac{\sin \alpha \, d\alpha}{\sin \theta \, d\theta}} = \frac{1 + e}{1 - e} \sqrt{\frac{(a - ez)d\alpha}{(a + ez)d\theta}} = \frac{1 + e}{1 - e} \frac{(a - ez)}{(a + ez)}. \tag{3.10}$$

Formula (3.10) is the analytical expression of the apodization factor of the elliptical mirror, and is continuous within the aperture angle interval of $[0, 2\pi]$.

In addition, by further analyzing the influence of the change in eccentricity of the elliptical mirror on its apodization factor, the result is shown in figure 3.3. The figure shows the change curves of apodization factors of elliptical mirrors with different eccentricities at different focusing angles, and compares the results to the curve of apodization factor of a perfect lens. On one hand, unlike the rule with respect to the lens that the apodization factor decreases as the focusing angle increases, the apodization factor of the elliptical mirror increases when its focusing angle increases, and the apodization factor of the elliptical mirror with a larger eccentricity grows faster. Since the scale of the apodization factor describes the degree of energy focusing of the transmitted light, the apodization factor that is greater than 1 shows that the energy is focused, and the apodization factor that is less than 1 shows that the energy is divergent. It can be concluded that the elliptical mirror can effectively focus the energy, especially in the region with a high aperture angle. On the other hand, different eccentricities of the ellipsoid correspond to the different maximum working aperture angles, for example, $e = 0.6$ of an elliptical mirror corresponds to a maximum aperture angle of $4\pi/3$, and when the eccentricity e reaches 0.8, the maximum aperture angle can be $5\pi/3$.

3.2 Vector focus model of elliptical mirror with extra high aperture angle

3.2.1 Analysis of focusing characteristic of elliptical mirror under circularly polarized illumination

Depending on the phase difference of components E_x and E_y, the circularly polarized light can be divided into right rotation circularly polarized light and left rotation circularly polarized light, which respectively satisfy:

$$E_y/E_x = -i,$$
$$E_y/E_x = i. \tag{3.11}$$

The advantage of focusing circularly polarized light is that the transversal light intensity distribution of the obtained focusing spot still has circular symmetry. Before specific discussion and without loss of generality, we assume that the incident light is left rotation circularly polarized light. The electric field near the focal point F_2 of the elliptical mirror is expressed as:

$$\begin{cases} E_x = \dfrac{iA}{\sqrt{2}}[I_0 + I_2 \cos (2\varphi_s) + iI_2 \sin(2\varphi_s)], \\[2mm] E_y = \dfrac{iA}{\sqrt{2}}[I_2 \sin (2\varphi_s) - iI_2 \cos (2\varphi_s) + iI_0], \\[2mm] E_z = -\sqrt{2}\, A[I_1 \cos \varphi_s + iI_1 \sin \varphi_s]. \end{cases} \tag{3.12}$$

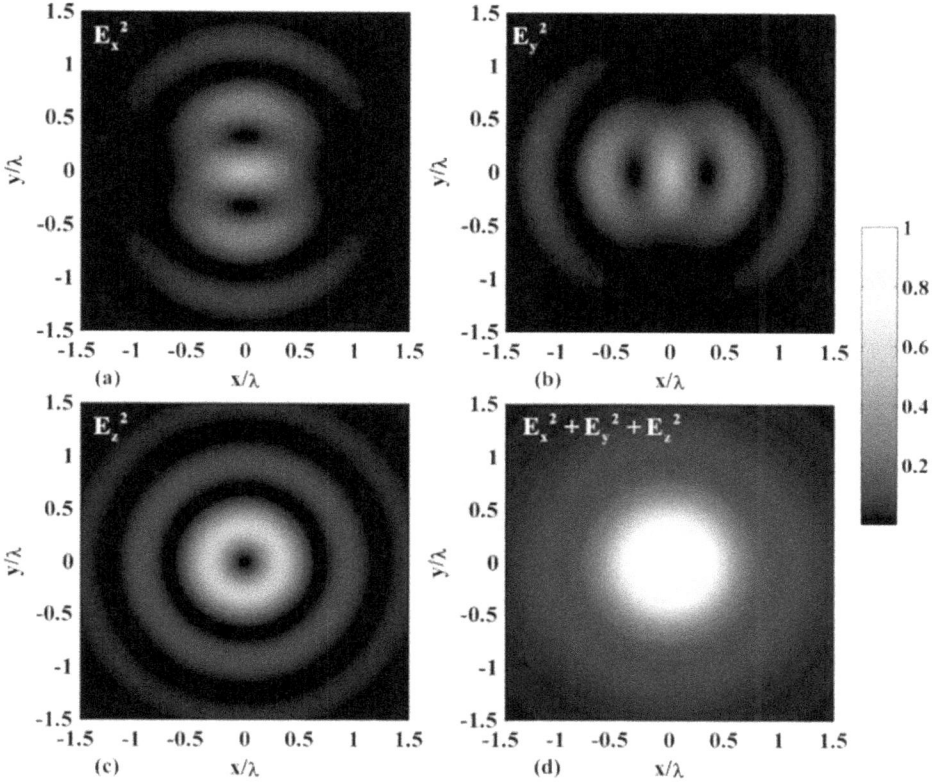

Figure 3.4. Transversal distribution of normalized electric field energy density when $e = 0.8$ and $\theta_{max} = 3\pi/4$ under circularly polarized illumination. (a) Electric field component in the X direction. (b) Electric field component in the Y direction. (c) Electric field component in the Z direction. (d) Total field.

Figure 3.4 shows the intensity distribution of the electric field on the focal plane under the circularly polarized illumination when the elliptical mirror has an eccentricity of 0.8 and an aperture angle of $3\pi/2$. It is worth noting that X and Y electric field components are distributed in a similar but mutually perpendicular way, and Z electric field component is annularly distributed. Unlike the focusing electric field under the linearly polarized illumination, three electric field components obtained by focusing circularly polarized light have a similar intensity, as shown in figure 3.5, the transversal half-width of focusing spot is about 1λ, which is much larger than the optical diffraction limit given by the Abbe diffraction limit definition. Moreover, the side lobe of the transversal point spread function is highly significant, which is caused by the strong side lobes of X and Y electric field components.

Figure 3.6 shows the axial characteristics of the focusing spot under the same circumstances. Since the aperture angle of the elliptical mirror exceeds π, the light spot continues to be compressed in the axial direction. Through comparison between the transversal and axial point spread functions (PSFs) of this time, we can find out that the width of the main lobe of the axial point spread function is much less than

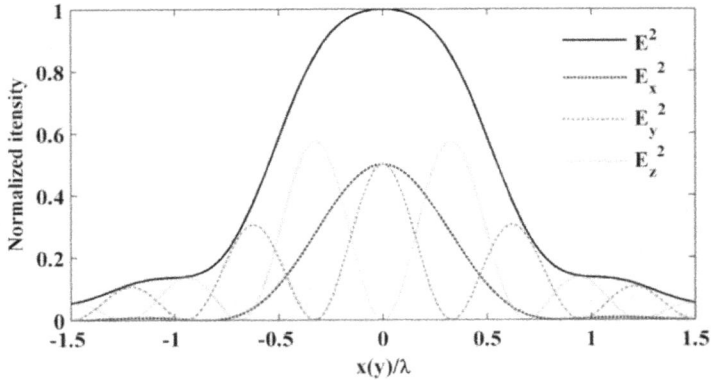

Figure 3.5. Normalized curves of various electric field components.

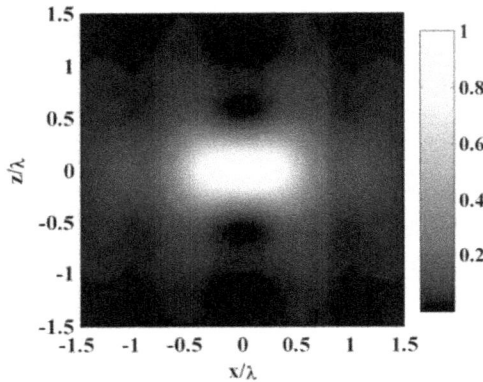

Figure 3.6. Axial distribution of normalize electric field energy density when $e = 0.8$ and $\theta_{max} = 3\pi/4$ under circularly polarized illumination.

that of the transversal point spread function (as shown in figure 3.7), and the side lobe of the former is also very small.

It is known that the optical diffraction limit in the axial direction is [6–8]:

$$R_z = 2\lambda/NA^2. \tag{3.13}$$

Obviously, for the focusing spot obtained by the elliptical mirror under the circularly polarized illumination, the axial half-width is decreased by about 3/4. In the industrial confocal measurement, a resolution that is much less than the axial size of light spot can be obtained by reading the axial envelope curve; however, the axial resolution of a bioconfocal microscope depends on the axial size of the focusing spot. Therefore, if an elliptical mirror with an extra-high aperture angle substitutes a traditional objective lens composed of lenses for bioconfocal microscopy imaging, the axial resolution can be improved.

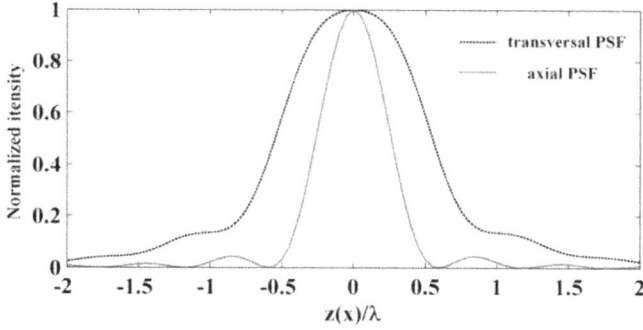

Figure 3.7. Comparison between transversal point spread function and axial point spread function.

3.2.2 Analysis of focusing characteristics of the elliptical mirror under radially polarized illumination

In addition to such uniform polarized light waves as linearly polarized light and circularly polarized light, modern optics theory derives research on the propagation and focusing characteristics of non-uniform polarized light. For example, radially polarized light is a form of non-uniform cylindrically symmetric vector beam. When a lens or parabolic mirror with a high numerical aperture is used to focus radially polarized light, a focusing spot smaller than the Abbe diffraction limit can be obtained. Meanwhile, electric field components in the vicinity of the focal point are special, and the principal component is the non-propagating axial electric field. At present, the radially polarized light has been widely used for the research on the generation of light needle, high-resolution microscopy, particle acceleration, etc [10, 11]. Considering the outstanding focusing characteristic of radially polarized light in a system with a high numerical aperture, this section will provide the vector focusing theory of radially polarized light in an elliptical mirror system with an extra-high aperture angle.

When an elliptical mirror is used to focus the radially polarized light, the electric field of focusing waves in the vicinity of the focal point F_2 can be expressed as:

$$\begin{cases} E_r(\rho_s, z_s) = A \int_0^{\theta_{max}} \sin 2\theta P(\theta) l(\alpha) J_1(k\rho_s \sin \theta) e^{-ikz_s \cos \theta} d\theta, \\ E_z(\rho_s, z_s) = -2iA \int_0^{\theta_{max}} \sin^2 \theta P(\theta) l(\alpha) J_0(k\rho_s \sin \theta) e^{-ikz_s \cos \theta} d\theta, \end{cases} \quad (3.14)$$

where, $E_r(\rho_s, z_s)$ is the radial electric field component, and $E_z(\rho_s, z_s)$ is the axial electric field component. $l(\alpha)$ is the distribution of waist of the Bessel-Gaussian beam on the front incident plane, and is specifically expressed as:

$$l(\alpha) = \exp[-\beta^2 (\sin \alpha / NA_{LENS})^2 J_1(2\beta_0 \sin \alpha / NA_{LENS})]. \quad (3.15)$$

β is the ratio of beam radius to beam waist, and generally, we can make $\beta = 1$. NA_{LENS} is the numerical aperture of front lens of the elliptical mirror system, and in the following discussion, we assume that $NA_{LENS} = 0.95$.

3-8

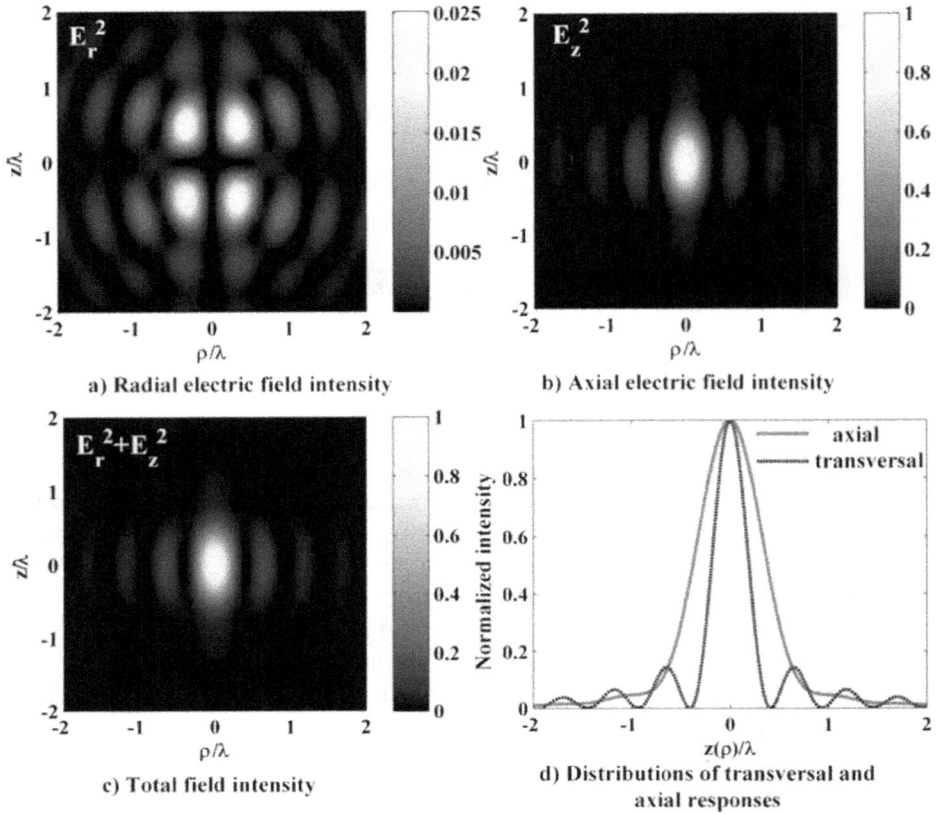

Figure 3.8. Distribution of normalized electric field energy density and point spread function curves when $e = 0.6$ and $\theta_{max} = 2\pi/3$ under radially polarized illumination. Adapted from [4]. (a) Radial electric field component. (b) Axial electric field component. (c) Total field. (d) Transversal and axial intensity distribution.

Similar to the above discussion, by assuming the elliptical mirror has an eccentricity of $e = 0.6$ and an aperture angle of $4\pi/3(240°)$, we can obtain the distribution of energy density of electric field in the vicinity of the focal point F_2, as shown in figure 3.8. Unlike the focusing characteristic when the aperture angle is smaller than π, the distribution of radial electric field components presents two independent rings surrounded by multiple side lobe rings. Meanwhile, energy of axial electric field components is concentrated in the vicinity of the focal point. Since the intensity of axial electric field components is much higher than that of the radial components, the distribution of total field is similar to that of the axial components. Figure 3.8(d) shows the distribution of transversal and axial intensity curves of the light spot. It can be seen that with further expansion of aperture angle of the elliptical mirror, both axial and transversal sizes of the light spot will be reduced. Although a transversal focusing spot has a larger side lobe compared with an Airy disk, this does not prevent the transversal focusing spot from being applied to the high-resolution microscopy imaging. However, if the aperture angle of elliptical mirror continues to increase, the size of focusing spot does not continue to decrease, and on the contrary, the size of the focusing spot increases, and the intensity of radial components also increases.

In order to clarify the condition under which the elliptical mirror system forms the smallest light spot under radially polarized illumination, table 3.1 shows elliptical mirrors with different eccentricities and aperture angles, and transversal and axial half-widths of their corresponding focusing spots. As seen from the table, for elliptical mirrors with different eccentricities, the axial half-widths of focusing spots decrease as the aperture angle increases, and the transversal half-widths decrease first and then increase as aperture angles increase. The elliptical mirror with a larger eccentricity can form a focusing spot with a smaller transversal half-width, but at the same aperture angle, the elliptical mirror with a smaller eccentricity can form a focusing spot with a smaller axial half-width.

Moreover, figure 3.9 shows the intensity ratios of radial to axial electric field components at different aperture angles when the elliptical mirror has an eccentricity of 0.6 and 0.8 respectively. In this figure, the black fold line shows the density ratio

Table 3.1. Half-width of transversal/axial point spread function of the focusing spot.

	$e = 0.6$			$e = 0.8$	
Half aperture angle (degree)	Transversal half-width (wavelength)	Axial half-width (wavelength)	Half aperture angle (degree)	Transversal half-width (wavelength)	Axial half-width (wavelength)
105	0.383	0.889	105	0.377	0.945
110	0.381	0.844	110	0.376	0.895
115	0.381	0.797	115	0.378	0.851
120	0.382	0.757	120	0.383	0.815

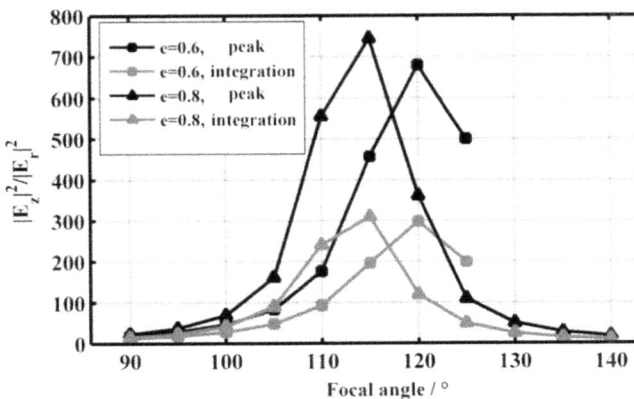

Figure 3.9. Intensity ratio of axial to radial electric field components. Adapted from [4].

of maximum electric field energy of axial to radial electric field components on the focal plane; the red fold line shows the energy ratio of axial to radial electric field components on the focal plane, and its reciprocal is the definition of polarization conversion efficiency of focusing light and is expressed as [9–11]:

$$\eta = \int_0^\infty |E_z(\rho, 0)|^2 2\pi\rho d\rho / (\int_0^\infty |E_r(\rho, 0)|^2 2\pi\rho d\rho + \int_0^\infty |E_z(\rho, 0)|^2 2\pi\rho d\rho). \quad (3.16)$$

As can be seen from the figure, the adoption of an elliptical mirror with an aperture angle exceeding π can bring up the polarization conversion efficiency of axial electric field of focusing light to more than 99% under appropriate circumstances. This discovery leads to the adoption of research findings of elliptical mirrors with high aperture angles in fields that require strong focusing spots of axial electric field components, such as particle acceleration, optical tweezers, etc. Moreover, a wave filter for the focusing system of an elliptical mirror is further designed in combination with the existing pupil filtering method, which not only improves the polarization conversion efficiency of focusing light, but also further reduces the lateral three-dimensional size of the focusing spot.

3.3 Conclusion

This chapter takes the Debye–Wolf vector diffraction integral as the basic model to build an elliptical mirror geometric equation expressed by the eccentricity, deduces an apodization factor analytical expression for elliptical mirrors with an aperture angle of 0 to 2π, and analyzes the focusing characteristics of the elliptical mirror system with an extra high aperture angle under circularly and radially polarized illumination. Under the radially polarized illumination, a focusing spot smaller than the diffraction limit can be generated by focusing, with transversal and axial half-widths of 0.38λ and 0.8λ respectively. The above theoretical analysis provides a basic theory for subsequent research on the confocal microscopy imaging system based on an elliptical mirror, and verifies the effectiveness of achieving high-resolution imaging by using an elliptical mirror with an extra-high aperture angle.

The resolution is a key factor of the microscopy imaging system, and is mainly influenced by the numerical aperture of objective lens and the properties of incident light. Therefore, the research on super-resolution focusing has developed into multiple approaches since 1950s, for example, the method of pupil filtering, the method of high numerical aperture polarized illumination, the method of surface plasma sub-wavelength focusing, and the method of negative refraction superlens focusing. This chapter discusses the method of using an elliptical mirror with a high numerical aperture to focus the polarized light, so as to improve the resolution. It is important to note that, although the base for theoretical research on the polarized light focusing method is complete and has endless stream of results, the prospect of practical application is not very clear, and this method has not attracted extensive attention in the field of practical application. One of the important reasons is that this technology cannot expand the cutoff frequency of an imaging system, and from the view of image quality, it improves the high frequency information imaging capability, and lowers the transmission capability of LF and IF signals. Therefore, one potential application of the technology is in a spot scanning microscopy imaging system (e.g. a confocal microscopic system based on an elliptical mirror) to achieve high-resolution imaging.

References

[1] Liu J, Tan J and Wilson T *et al* 2012 Rigorous theory on elliptical mirror focusing for point scanning microscopy *Opt. Express* **20** 6175–84

[2] Liu J, Zhong C and Tan J *et al* 2012 Elliptical mirror based imaging with aperture angle greater than π/2 *Opt. Express* **20** 19206–13

[3] Ji L, Min A, Jiubin T, Rui W and Xinran T 2013 Focusing of cylindrical-vector beams in elliptical mirror based system with high numerical aperture *Opt. Commun.* **305** 71–5

[4] Liu J, Ai M and Zhang H *et al* 2013 Focusing properties of elliptical mirror with an aperture angle greater than π *Opt. Eng.* **53** 061606

[5] Liu J, Ai M and Zhang H *et al* 2013 Focusing of an elliptical mirror based system with aberrations *J. Opt.* **15** 105709

[6] Wilson T 1990 *Confocal Microscopy* (London: Academic)

[7] Sheppard C J R and Wilson 1981 The theory of the direct-view confocal microscope *J. Microsc.* **124** 107–17

[8] Sheppard C J R and Gu M 1991 Three-dimensional optical transfer function for an annular lens *Opt. Commun.* **81** 276–80

[9] Wang H, Shi L and Lukyanchuk B *et al* 2008 Creation of a needle of longitudinally polarized light in vacuum using binary optics *Nat. Photonics* **2** 501–505

[10] Zhan Q 2004 Trapping metallic Rayleigh particles with radial polarization *Opt. Express* **12** 3377–82

[11] Lin J, Yin K and Li Y *et al* 2011 Achievement of longitudinally polarized focusing with long focal depth by amplitude modulation *Opt. Lett.* **36** 1185–7

IOP Publishing

Elliptical Mirrors
Applications in microscopy
Jian Liu

Chapter 4

Imaging analysis of dipole vector in an elliptical mirror

Yuhang Wang, Jian Liu, Cien Zhong and Jiubin Tan

4.1 Introduction

In previous chapters, focusing properties of different polarized lights in elliptical mirror have been respectively analyzed, including linearly polarized light, circularly polarized light and radially polarized light. In addition, from the applied perspective, to observe molecule microstructure via an optical microscope based on an elliptical mirror is also an important bioresearch approach. So in this chapter, imaging analysis of a single molecule that models as a dipole vector in an elliptical mirror will be introduced.

In recent years, lots of researchers have introduced their findings into the area of single molecule imaging. With the development of new techniques for microscopy, the observation of molecules in the cryogenic solid has been performed [1]. The technology, in which a near-field optical microscope is used to determine space direction of single molecular dipole moment, is also developed [2]. In addition, the angular direction of the dipole can be derived according to the intensity distribution of the single molecule fluorescence in an objective image field. This method has been applied for the determination of the dipole moment direction of a single man-made fiber molecule [3]. A fluid-immersion objective with high numerical aperture is used to observe the intensity distribution of image field of an excited dipole [4]. The author of this paper first discovered that the dipole in the vibration direction perpendicular to optical axis is focused, while the dipole in the vibration direction parallel to the optical axis is out of focus and its focal spot is in the form of a donut. A telecentric optical system, consisted of a parabolic mirror with high numerical aperture and an aplanatic lens with low numerical aperture, is used to detect those electric dipole emitters placed at a uniform environment [5]. Its analytical results show that the spatial vibration direction of the electric dipoles can be determined according to the imaging characteristics of a single electric dipole. Meanwhile, in the research of molecule imaging, the image quality can be improved by collecting more exciting light, instead of purely increasing the light intensity exposed to samples.

This chapter will discuss and analyze how to detect the image field distribution of the electric dipole emitter via a telecentric optical system consisted of an aplanatic objective with high numerical aperture and an aplanatic objective with low numerical aperture. The imaging characteristics of a single electric dipole in an elliptical mirror with high numerical aperture will also be studied. The details are as follows. In section 4.2, the three-dimensional expression for the focus region of an electric dipole emitter in an elliptical mirror and an aberration free dual-lens system will be derived on the basis of the Debye–Wolf vector diffraction theory. In sections 4.3 and 4.4, the imaging characteristics of the dipole in the focal point will be analyzed, the energy density distribution diagram for focus region of light waves excited by electric dipoles in the elliptical mirror or aplanatic double lens system will be given, the energy density distribution law of the focus region where the dipole is placed away from the focal point will be analyzed, and those energy density distribution laws of the focused field where the electric dipole is located in two kinds of system in both cases of lateral defocusing and axial defocusing will be given. Finally, in section 4.5, we will analyze the comparison diagram for energy density envelope curves of focused electric fields of coordinate axes if the dipole is located in the focal point and if it is focused to an imaged in an elliptical mirror, a parabolic mirror and an aplanatic dual lens.

4.2 Imaging model of dipole vector in elliptical mirror

In typical electromagnetic field theories, a single molecule fluorescence excitation is considered as the radiation of electric dipole formed in internal molecule processes. An electric dipole is a system composed of two closely spaced point charges of equal energy and opposite sign, with the distance l between positive and negative charge centers, and the direction towards positive charge center from negative charge center. The electric dipole is characterized by the electric dipole moment p.

$$p = ql. \tag{4.1}$$

The vibrating electric dipole will produce an alternative electromagnetic field around its space, which means radiating out light waves. The electric field radiated by the electric dipole in a distant field can be expressed as

$$E_{de} = \frac{k^2}{\sqrt{4\pi\varepsilon_0}}(s_m \times p) \times s_m \cdot \frac{1}{r} \cdot \exp{(ikr - wt)} \tag{4.2}$$

where, ε_0 is a dielectric constant in vacuum, k is the number of waves, p is a dipole moment, $\mathbf{S}_m$ is a unit vector in the observation direction, r is the distance from dipole to observation point, w is an angular frequency, and t is the time.

The geometric diagram for the imaging of electric dipole in elliptical mirror system is shown in figure 4.1. The electric dipole is placed at the right focal point F_1 of the elliptical mirror, a photo detector is placed at the left focal point F_2 thereof, and the point of origin is based in point O, i.e., the center of ellipsoid. The light

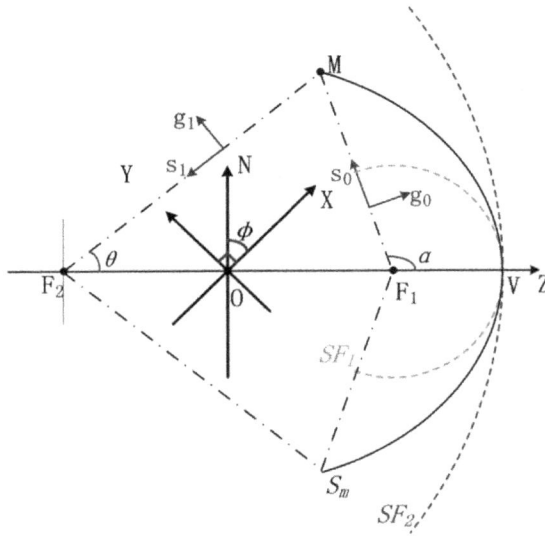

Figure 4.1. The geometrical relationship of vectors when the electric dipole images in the elliptical mirror [6].

waves radiated by the electric dipole emitter will be focused to the left focal point F_2 of the elliptical mirror after convergence. If the photo detector or other light intensity detecting device is placed at the focal point F_2, an imaging spot of the electric dipole in the elliptical mirror will be obtained. As shown in figure 4.1, SF_1 is a Gaussian spherical focal surface formed by the incident light on the elliptical mirror, with the radius of $a–c$, where a is the major axis of the elliptical mirror, and c is the centrifugal distance thereof. SF_2 is the Gaussian spherical focal surface formed by the reflected light on the elliptical mirror, with the radius of $a + c$. S_m is the surface of the elliptical mirror. As the light waves are transverse, their electric vectors can be formed by superimposing two vertical components perpendicular to the propagation direction. The two orthogonal vectors perpendicular to the propagation direction within an object space are g_0 and s_0 respectively, while the two orthogonal vectors perpendicular to the propagation direction within an image space are g_1 and s_1 respectively. The aperture angles of object space and of image space are α and θ respectively. The angle between the meridian plane and the axis x is Φ. The vectors g_0, s_0, g_1 and s_1 can be expressed as

$$s_0 = \sin \alpha \cos \phi \mathbf{i} + \sin \alpha \sin \phi \mathbf{j} + \cos \alpha \mathbf{k} \qquad (4.3)$$

$$g_0 = -\cos \alpha \cos \phi \mathbf{i} - \cos \alpha \sin \phi \mathbf{j} + \sin \alpha \mathbf{k} \qquad (4.4)$$

$$s_1 = -\sin \theta \cos \phi \mathbf{i} - \sin \theta \sin \phi \mathbf{j} - \cos \theta \mathbf{k} \qquad (4.5)$$

$$g_1 = \cos \theta \cos \phi \mathbf{i} + \cos \theta \sin \phi \mathbf{j} - \sin \theta \mathbf{k} \qquad (4.6)$$

where, i, j, and k are unit vectors on coordinate axes x, y and z respectively.

Let the electric field at the exit pupil after being reflected by the elliptical mirror be e_1, within a plane consisted of vectors g_1 and $g_1 \times s_1$, that is

$$e_1 = \beta g_1 + \gamma(g_1 \times s_1). \tag{4.7}$$

According to the principle that radial components are unchanged and azimuthal components are reversed when light waves pass through a mirror surface, then

$$\beta = e_1 \cdot g_1 = E_{de} \cdot g_0 \tag{4.8}$$

$$\gamma = e_1 \cdot (g_1 \times s_1) = -E_{de} \cdot (g_0 \times s_0). \tag{4.9}$$

For simplicity, it is assumed that the energy loss caused by elliptical reflection and absorption can be neglected, then using Wolf diffraction integral, the electric field adjacent to the left focal point F_2 (independent of time) can be expressed as follows:

$$\left.\begin{aligned}
E_{fe} &= \frac{-ik^3}{\sqrt{16\pi^3\varepsilon_0}} \frac{a+c}{a-c} \int_{\theta_{\min}}^{\theta_{\max}} \int_0^{2\pi} l(\theta)\{[(\boldsymbol{p} \cdot \boldsymbol{g}_0)\boldsymbol{g}_1 - [\boldsymbol{p} \cdot (\boldsymbol{g}_0 \times \boldsymbol{s}_0)](\boldsymbol{g}_1 \times \boldsymbol{s}_1)]\} \\
&\quad \times \exp[ik(r_p s_1 - n_m r_m s_0)]\sin\theta d\theta d\phi \\
l(\theta) &= \frac{a-c}{a+c}\sqrt{\frac{\sin\alpha}{\sin\theta} \cdot \frac{d\alpha}{d\theta}}
\end{aligned}\right\} \tag{4.10}$$

where, a is the major axis of the elliptical mirror, and c is the centrifugal distance thereof. r_p is the location vector of the probe point, as follows:

$$r_p = r_p \sin\theta_p \cos\phi_p \boldsymbol{i} + r_p \sin\theta_p \sin\phi_p \boldsymbol{j} + r_p \cos\theta_p \boldsymbol{k}. \tag{4.11}$$

r_m is the distance vector of the dipole directed by focal point F_1, n_m is the refractive index of medium, and $l(\theta)$ is the apodization factor of the elliptical mirror. The upper and lower limits of the image space focusing angle θ are as follows:

$$\left.\begin{aligned}
\theta_{\min} &= \arctan(r_{samp}/2c), \\
\theta_{\max} &= \arctan\left(\frac{z_m - c}{z_m + c} \tan(\arcsin(NA_m/n_m))\right)
\end{aligned}\right\} \tag{4.12}$$

where, r_{samp} is the radius of center shielded by a sample supporting rack, NA_m is the numerical aperture of ellipsoid, n_m is the refractive index of medium, and z_m is coordinate z of points on the ellipsoid if the aperture angle is maximum. If $\alpha \in [0, \pi/2]$, the relationship between image space and object space aperture angles is as follows:

$$\alpha = \arctan[(z + c)\tan\theta/(z - c)]. \tag{4.13}$$

If $\alpha \in [\pi/2, \pi]$, the relationship between image space and object space aperture angles is as follows:

$$\alpha = \pi - \arctan[(z + c)\tan\theta/(c - z)] \tag{4.14}$$

where, z, described in the above two expressions, is component z of any point M on the ellipsoid, and can be obtained according to the geometrical relationship, as follows:

$$z = \left(ab^2\sqrt{1 + \tan^2\theta} - a^2c\tan^2\theta\right)/(a^2\tan^2\theta + b^2). \qquad (4.15)$$

The mathematical calculation model for dipole imaging in both cases of dipole moment $p = i$ and $p = k$ will be discussed below. If the dipole is placed at the focal point F_1, and the dipole moment $p = i$, the electric field at the detected end can be reduced to

$$\left.\begin{aligned}
E_x &= iA(I_0 + I_2\cos 2\phi_p), \\
E_y &= iAI_2\sin 2\phi_p, \\
E_z &= -2AI_1\cos\phi_p.
\end{aligned}\right\} \qquad (4.16)$$

where, constant term $A = \dfrac{\pi k^3}{\sqrt{16\pi^3\varepsilon_0}}$, and integral term is as follows:

$$\left.\begin{aligned}
I_0 &= \int_0^{\theta_{max}}\sqrt{\frac{\sin\alpha\,d\alpha}{\sin\theta\,d\theta}}(1 + \cos\theta\cos\alpha)\sin\theta J_0(kr_p\sin_p\sin\theta)e^{-ikz_s\cos\theta}d\theta, \\
I_1 &= \int_0^{\theta_{max}}\sqrt{\frac{\sin\alpha\,d\alpha}{\sin\theta\,d\theta}}\sin^2\theta\cos\alpha J_1(kr_p\sin_p\sin\theta)e^{-ikz_s\cos\theta}d\theta, \\
I_2 &= \int_0^{\theta_{max}}\sqrt{\frac{\sin\alpha\,d\alpha}{\sin\theta\,d\theta}}(1 - \cos\theta\cos\alpha)\sin\theta J_2(kr_p\sin_p\sin\theta)e^{-ikz_s\cos\theta}d\theta.
\end{aligned}\right\} \qquad (4.17)$$

Formulas (4.16) and (4.17) show the three-dimensional expressions of the focused electric field after being reflected by the elliptical mirror if the electric dipole vibrating along the direction x is placed at the focal point of the elliptical mirror.

If the dipole moment $p = k$ and other two conditions are unchanged, the electric field at the detected end can be reduced to

$$\left.\begin{aligned}
E_x &= -2AI_1\cos\phi_p, \\
E_y &= -2AI_1\sin\phi_p, \\
E_z &= 2iAI_0,
\end{aligned}\right\} \qquad (4.18)$$

where, the constant term A is consistent with the previous value, and the integral term changes to

$$\left.\begin{aligned}
I_0 &= \int_0^{\theta_{max}}\sqrt{\frac{\sin\alpha\,d\alpha}{\sin\theta\,d\theta}}\sin^2\theta\sin\alpha J_0(kr_p\sin_p\sin\theta)e^{-ikz_s\cos\theta}d\theta, \\
I_1 &= \int_0^{\theta_{max}}\sqrt{\frac{\sin\alpha\,d\alpha}{\sin\theta\,d\theta}}\sin\theta\cos\theta\sin\alpha J_1(kr_p\sin_p\sin\theta)e^{-ikz_s\cos\theta}d\theta.
\end{aligned}\right\} \qquad (4.19)$$

Formulas (4.18) and (4.19) show the three-dimensional expressions of the focused electric field after being reflected by the elliptical mirror if the electric dipole vibrating along the direction z is placed at the focal point of the elliptical mirror.

4.3 Imaging characteristics of the electric dipole in the elliptical mirror

Formulas (4.16) and (4.17) and formulas (4.18) and (4.19) respectively show the three-dimensional expressions of the focused electric field after being reflected by the elliptical mirror if the electric dipole vibrating along the direction x or z is placed at the focal point of the elliptical mirror. It is assumed that the major axis a of the elliptical mirror is 500 mm in length, and the minor axis b thereof is 400 mm in length. For the elliptical mirror, its numerical aperture is 1.515, the refractive index of medium is 1.518, and the radius of central obscuration caused by the sample bracket is 10 mm. If the dipole is placed at the right focal point F_1 of the elliptical mirror, the image of $|E|^2$ detected by the photo detector at the left focal point F_2 is as shown in figure 4.2. Subgraph (a) shows the dipole moment of electric dipole, $p = i$; subgraph (b) shows the dipole moment of electric dipole, $p = j$; and subgraph (c) shows the dipole moment of electric dipole, $p = k$. Those red lines shown in the figures are Airy disk profiles.

In figure 4.2, we can observe that the vibration direction of the dipole moment of electric dipole has a direct influence on the shape of spot obtained at the detected end. As shown in figures 4.2(a) and (b), if the dipole vibrates in a lateral direction (direction x or y), the spot center obtained is solid. As shown in figure 4.2(c), if the dipole vibrates in an axial direction (direction z), the focal imaging spot of which the center is dark may be obtained. The vibration direction of the electric dipole can be determined according to dipole imaging laws.

If the dipole is placed at a position, which is one wavelength away from the right focal point F_1 along the focal plane of the ellipsoidal mirror in the direction forming a $30°$ angle with the axis x, then

$$r_m = \frac{\lambda}{n_m}\left[\cos\left(\frac{\pi}{6}\right), \ \sin\left(\frac{\pi}{6}\right), \ 0\right]. \tag{4.20}$$

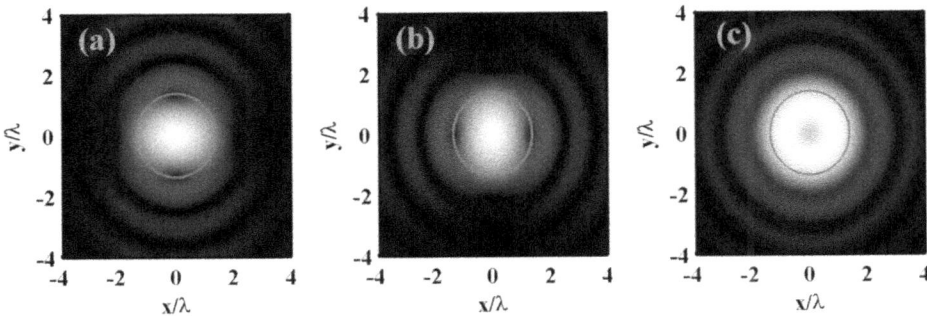

Figure 4.2. Focal plane profile for energy density of a focused electric field of an electric dipole placed at the focal point of the elliptical mirror.

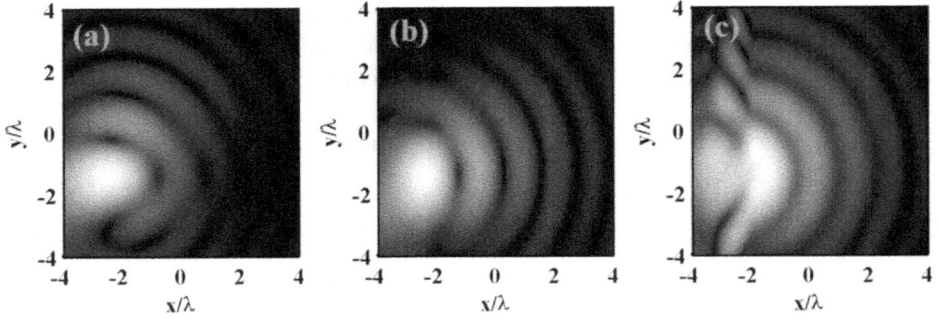

Figure 4.3. Focal plane profile for energy density of a focused electric field of the electric dipole placed in the elliptical mirror along the lateral direction away from the focal point.

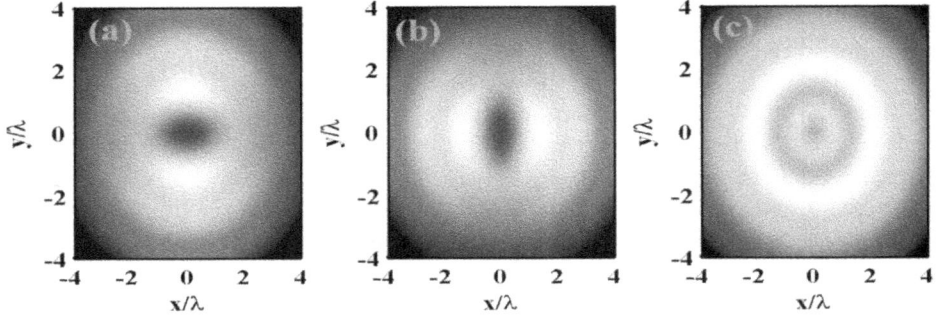

Figure 4.4. Focal plane profile for energy density of the focused electric field of the electric dipole placed in the elliptical mirror along the central axis away from the focal point.

The energy density distribution diagram of the focused electric field obtained at the left focal point F_2 is as shown in figure 4.3. Subgraph (a) shows the dipole moment of electric dipole, $p = i$; subgraph (b) shows the dipole moment of electric dipole, $p = j$; and subgraph (c) shows the dipole moment of electric dipole, $p = k$.

In figure 4.3, we can observe that coma will occur in the spot detected at the image space focal plane if the dipole is placed at a position, which is one wavelength away from the focal point along the plane x–y. And like the imaging laws of electric dipole in parabolic mirror, the elliptical mirror is sensitive to an off-axis aberration.

In the elliptical mirror, if the electric dipole is placed at a position, which is one wavelength away from the focal point along axis z, then

$$r_m = \lambda(0,\ 0,\ 1)/n_m. \tag{4.21}$$

The electric dipole image obtained at the left focal point F_2 is as shown in figure 4.4. Subgraph (a) shows the dipole moment of electric dipole, $p = i$; subgraph (b) shows the dipole moment of electric dipole, $p = j$; and subgraph (c) shows the dipole moment of electric dipole, $p = k$.

In figure 4.4, we can observe that the spot center measured at the detected end is dark, if the electric dipole is placed at a position, which is one wavelength away from

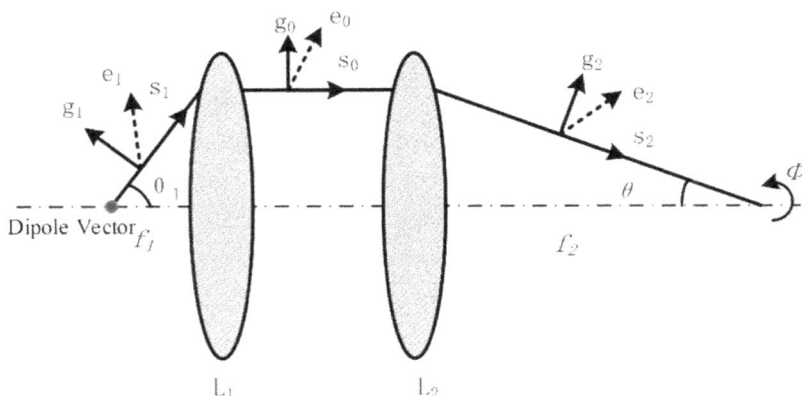

Figure 4.5. The geometrical relationship of vectors when the electric dipole images in a dual-lens system.

the focal point along axis z. This is similar to the focal plane profile for energy density of focused electric field of electric dipole vibrating in the direction z.

4.4 Imaging characteristics of the electric dipole in a dual-lens system

We have already known that imaging characteristics of electric dipole in elliptical mirror is sensitive to off-axis aberration from the above analysis. Does it mean that its characteristics in dual-lens system follow the same principle? Of course not; the analysis will be described in more detail in the following content.

Firstly, a simplified dual-lens system for the imaging of an electric dipole is shown in figure 4.5. The electric dipole is placed at the front focal point of L_1 lens and the detector is located at the back focal point of L_2 lens in the dual-lens system. Aperture angles in objective field and imaging field are respectively represented by θ_1 and θ. Focal lengths of L_1 and L_2 lens are represented by f_1 and f_2 respectively. The intersection angle between the meridian plane and x axis is Φ. Their electric vectors can be formed by superimposing two vertical components perpendicular to the propagation direction. The two orthogonal vectors perpendicular to the propagation direction within an object space are g_0 and s_0 respectively, while the two orthogonal vectors perpendicular to the propagation direction within an image space are g_1 and s_1 respectively.

When the dipole is placed on the front focal point of L_1 lens, and the dipole moment p equals i, the expression of detective intensity of electrical field can be simplified as follows:

$$
\left.
\begin{aligned}
E_x &= -iA(I_0 + I_2 \cos 2\phi_p), \\
E_y &= -iAI_2 \sin 2\phi_p, \\
E_z &= -2AI_1 \cos \phi_p.
\end{aligned}
\right\} \tag{4.22}
$$

where, constant term is $A = \dfrac{k^3 f_2}{f_1 \sqrt{16\pi\varepsilon_0}}$, and integral term is shown as follows:

$$\left.\begin{aligned}
I_0 &= \int_0^{\theta_{max}} \sqrt{\frac{\cos\theta}{\cos\theta_1}}(1 + \cos\theta\cos\theta_1)\sin\theta J_0(kr_p\sin_p\sin\theta)e^{ikz_s\cos\theta}d\theta, \\
I_1 &= \int_0^{\theta_{max}} \sqrt{\frac{\cos\theta}{\cos\theta_1}}\sin^2\theta\cos\theta_1 J_1(kr_p\sin_p\sin\theta)e^{-ikz_s\cos\theta}d\theta, \\
I_2 &= \int_0^{\theta_{max}} \sqrt{\frac{\cos\theta}{\cos\theta_1}}(1 - \cos\theta\cos\theta_1)\sin\theta J_2(kr_p\sin_p\sin\theta)e^{-ikz_s\cos\theta}d\theta.
\end{aligned}\right\} \quad (4.23)$$

Formulas (4.22) and (4.23) show the three-dimensional expressions of the focused electric field in the aberration free dual-lens system, when the electric dipole vibrates along the direction x.

If the dipole moment $p = k$ and other two conditions are unchanged, the electric field at the detected end can be reduced to

$$\left.\begin{aligned}
E_x &= 2AI_1\cos\phi_p, \\
E_y &= 2AI_1\sin\phi_p, \\
E_z &= 2iAI_0,
\end{aligned}\right\} \quad (4.24)$$

where, constant term $A = \dfrac{\pi k^3}{\sqrt{16\pi^3\varepsilon_0}}$, and integral term is as follows:

$$\left.\begin{aligned}
I_0 &= \int_0^{\theta_{max}} \sqrt{\frac{\cos\theta}{\cos\theta_1}}\sin^2\theta\sin\theta_1 J_0(kr_p\sin_p\sin\theta)e^{ikz_s\cos\theta}d\theta, \\
I_1 &= \int_0^{\theta_{max}} \sqrt{\frac{\cos\theta}{\cos\theta_1}}\sin\theta\cos\theta\sin\theta_1 J_1(kr_p\sin_p\sin\theta)e^{ikz_s\cos\theta}d\theta.
\end{aligned}\right\} \quad (4.25)$$

Formulas (4.22)–(4.25) show the three-dimensional expressions of the focused electric field if the electric dipole is placed in the aplanatic double lens. If the focal distance (f_1), numerical aperture (NA_m) and refractive index (N_m) of the left lens L_1 are 25 mm, 1.515 and 1.518 respectively, and the focal distance (f_2) of the right lens L_2 is 100 mm, then the system magnification (M) is 4. If the dipole is placed at the focal point F_1 of the left lens L_1, the image of electric field energy density $|E|^2$ detected by the photo detector at the right focal point F_2 is as shown in figure 4.6. Subgraph (a) shows the dipole moment of electric dipole, $p = i$; subgraph (b) shows the dipole moment of electric dipole, $p = j$; and subgraph (c) shows the dipole moment of electric dipole, $p = k$. Those red lines shown in the figures are circles with the radius, the same as the Airy disk.

In figure 4.6, we can observe that the size of the main bright spot in the center is basically the same as that of the Airy disk. If the dipole is placed at the focal point,

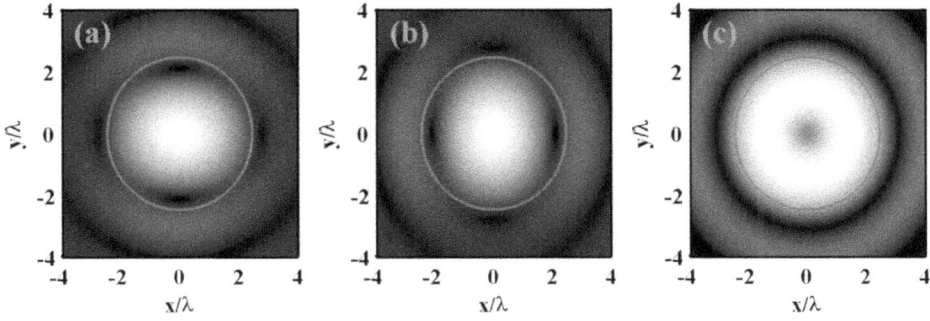

Figure 4.6. Focal plane profile for energy density of the focused electric field of the electric dipole placed at the focal point of an aplanatic double lens.

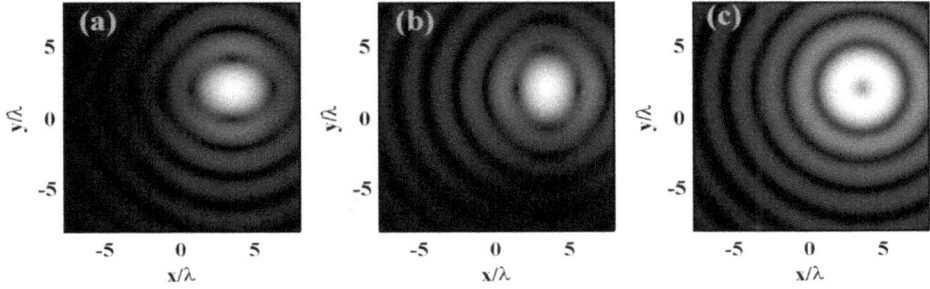

Figure 4.7. Imaging of the dipole placed in a dual-lens system along focal plane away from focal point.

the imaging spot shape of the focal plane at the detected end after passing through the dual-lens system is similar to that passed through the elliptical mirror system and the parabolic mirror system. But, because the dual-lens system has a small image space numerical aperture, the resultant Airy disk radius is large. If the dipole vibrates in a lateral direction (direction x or y), the spot center obtained is solid, and if the dipole vibrates in an axial direction (direction z), the spot center obtained is dark, which is a universal law of the dual-lens system, parabolic mirror system and elliptical mirror system.

If the dipole is placed at a position, which is one wavelength away from the right focal point F_1 along the focal plane of the aplanatic double lens in the direction forming a 30° angle with axis x, then

$$r_m = \frac{\lambda}{n_m}\left[\cos\left(\frac{\pi}{6}\right), \ \sin\left(\frac{\pi}{6}\right), \ 0 \right]. \tag{4.26}$$

Figure 4.7 shows the profile for energy density $|E|^2$ of the focused electric field detected at the right focal point F_2 by using the photo detector. Subgraph (a) shows the dipole moment of electric dipole, $p = i$; subgraph (b) shows the dipole moment of electric dipole, $p = j$; and subgraph (c) shows the dipole moment of electric dipole, $p = k$.

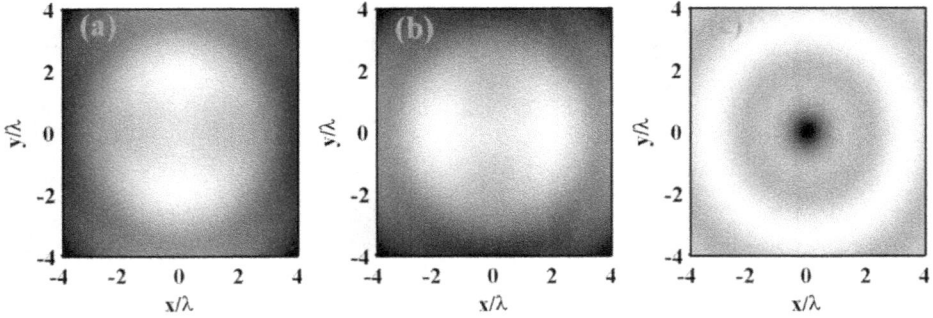

Figure 4.8. Imaging of the dipole placed in a dual-lens system along axis Z away from the focal point.

In figure 4.7, we can observe that there has no coma in the profile for energy density of the focused electric field obtained at the detected end, compared with the parabolic mirror system and elliptical mirror system, if the dipole images along the focal plane away from the focal point in the dual-lens system. The simulation results indicate that the lens system is insensitive to off-axis aberration.

Figure 4.8 shows the profile for field energy density $|E|^2$ detected from the focal plane at the back focal point F_2 of the lens L_2 by using the photo detector, if the electric dipole is placed at a position which is one wavelength away from the focal point along axis z of the aplanatic double lens. Subgraph (a) shows the dipole moment of electric dipole, $p = i$; subgraph (b) shows the dipole moment of electric dipole, $p = j$; and subgraph (c) shows the dipole moment of electric dipole, $p = k$. In figure 4.8, we can observe that the focal plane profile for energy density of the focused electric field obtained at the detected end is similar to that of electric dipole vibrating along the direction z placed in the focal point when the dipole images in the dual-lens system, if the dipole is placed at a position which is one wavelength away from the focal point along the axis z. In figures 4.8(a) and (b), we can observe that the central spot is enlarged, and the brightness weakens substantially compared with the case where the dipole is placed in the focal point, if the electric dipole vibrates laterally (along the direction x or y). In figure 4.8(c), we can observe that the spot center is dark and has a large radius, compared with the case where the dipole is placed in the focal point, if the electric dipole vibrates axially (along the direction z).

4.5 Comparison on imaging characteristics of dipole in elliptical mirror, parabolic mirror and lens

If the elliptical mirror with an aperture angle greater than $\pi/2$ is used to perform the focal imaging on the electric dipole, not only the forward diffraction field of exciting light can be collected, but also part of the backward diffraction field can be collected. Figure 4.9 shows the envelope curves of focused fields of electric dipole in elliptical mirror with aperture angles of $\pi/2$ and $2\pi/3$ and in parabolic mirror with aperture angle of $\pi/2$.

As shown in this figure, the main lobe width of the lateral envelope curve, if the electric dipole is placed in the focused field of the elliptical mirror with an aperture

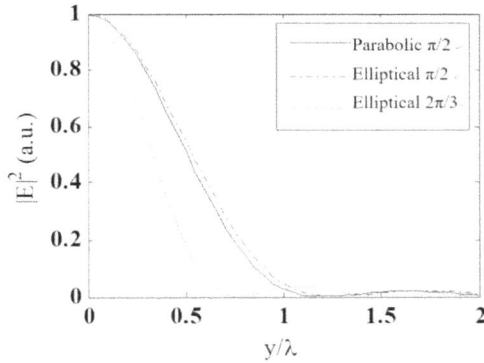

Figure 4.9. Envelope curve of focused field of electric dipole.

angle of $2\pi/3$, is the narrowest, which indicates that its image resolution is the highest, and accordingly, its image quality is the best. This is again a proof that the image quality of a fluorescence microscope can be improved by collecting more exciting light, as mentioned in James B Pawley' monograph. Because if the elliptical mirror with an aperture angle greater than $\pi/2$ is used to perform focal imaging in electric dipole, not only the forward diffraction field of reflected light of electric dipole can be collected, but also part of backward diffraction field can be collected, collecting more photons means that more information about samples is obtained, and its final image quality is also improved.

The imaging characteristics of electric dipole in dual-lens system, parabolic mirror system and elliptical mirror system not only have some similarities, but also have significant differences. During numerical simulation, it is assumed that all the object space numerical apertures of these three systems are 1.515, the refractive indexes of medium are 1.518 and the system magnifications are 4, and that like in the dual-lens system, there are no shelters for a sample supporting rack in the parabolic mirror system and elliptical mirror system.

If the electric dipole vibrating along the direction x is placed at the focal point of each system, the lateral and axial variation curves of the energy density for focused electric field of each system are as shown in figure 4.10. Figure 4.10(a) shows the variation curve of axis x, and figure 4.10(b) shows the variation curve of axis z.

In figure 4.10, we can observe that the lateral and axial resolutions of the focal spot are high under the identical condition, if the dipole is placed in the second order aspheric mirror.

If the electric dipole vibrates along the direction z, the lateral and axial variation curves of the energy density for focused electric field are as shown in figure 4.11. Figure 4.11(a) shows the variation curve of axis x, and figure 4.11(b) shows the variation curve of axis z. In figure 4.11, we can observe that the lateral and axial resolutions of the energy density for focused electric field of the electric dipole in the second order aspheric mirror are high, and the value and size of the dark space in the center of the lateral envelope are small under the identical condition, if the dipole vibrates along the direction z.

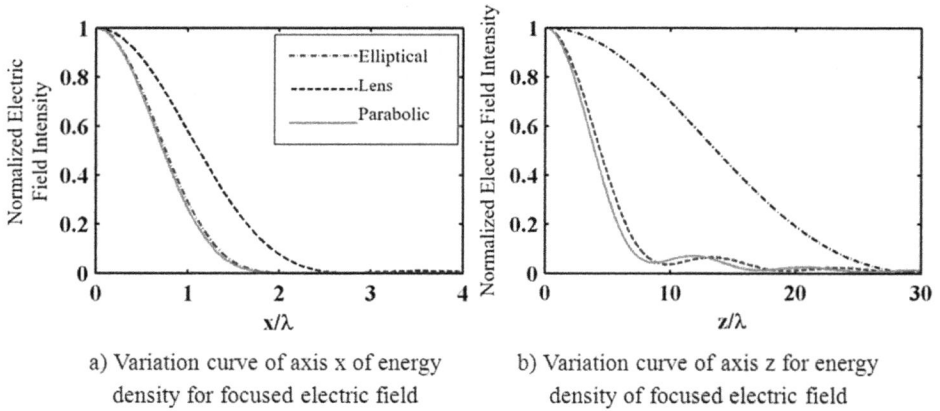

a) Variation curve of axis x of energy
density for focused electric field

b) Variation curve of axis z for energy
density of focused electric field

Figure 4.10. The energy density curve of a focused electric field of the electric dipole vibrating along direction x in three systems.

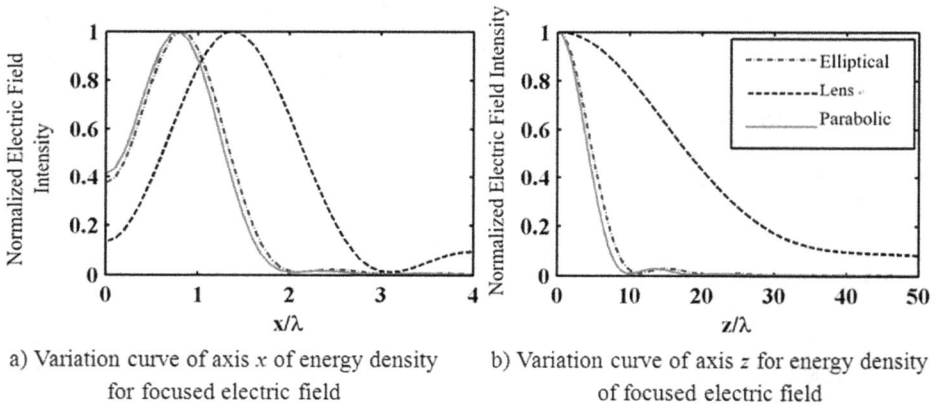

a) Variation curve of axis x of energy density
for focused electric field

b) Variation curve of axis z for energy density
of focused electric field

Figure 4.11. The energy density curve of the focused electric field of the electric dipole vibrating along direction z in three systems.

4.6 Summary

Firstly, the three-dimensional representations of the focused electric field of an electric dipole emitter in an elliptical mirror and aplanatic double lens are derived on the basis of Debye–Wolf diffraction integral. Numerical simulation analysis is performed on the focal plane profile of energy density for focused electric field in the case that the vibration direction of the electric dipole is parallel to or perpendicular to the optical axis (axis z). The results show that the spatial vibration direction of the electric dipole can be determined according to the imaging law of electric dipole, if the electric dipole emitter is placed at the focal point. Besides, the numerical simulation analysis is performed on the focal plane profile of energy density for the focused electric field of an electric dipole placed away from axis. The simulation results show that the second-order aspheric mirror is more sensitive to off-axis aberration. Finally, the lateral and axial variation laws of the focused electric field of the electric dipole in the elliptical mirror, parabolic mirror and aplanatic double lens are compared. The simulation results show that the lateral and axial resolutions of the focal spot are high if the electric dipole is placed in the second-order aspheric mirror. If the elliptical mirror with an aperture angle greater than $\pi/2$ is used to perform focal imaging in electric dipole, not only the forward diffraction field of reflected light of electric dipole can be collected, but also part of backward diffraction field can be collected, thereby improving the image quality.

References

[1] Moerner W E, Plakhotnik T, Irngartinger T, Croci M, Palm V and Wild U P 1994 Optical probing of single molecules of terrylene in a Shpol'kii matrix: a two-state single-molecule switch *J. Phys. Chem.* **98** 7382–9
[2] Betzig E and Chichester R J 1993 Single molecules observed by near-field scanning optical microscopy *Science* **262** 1422–5
[3] Sepiol J, Jasny J, Keller J and Wild U P 1997 Single molecules observed by immersion mirror objective *J. Phys. Chem.* **273** 444–8
[4] Dickson R M, Norris D J and Moerner W E 1998 Simultaneous imaging of individual molecules aligned both parallel and perpendicular to the optic axis *Phys. Rev. Lett.* **81** 5322–5
[5] Lieb M A and Meixner A J 2001 A high numerical aperture parabolic mirror as imaging device for confocal microscopy *Opt. Express* **8** 458–74
[6] Liu J, Cien Z, Jiubin T, Tong W and Tony W 2012 Elliptical mirror based imaging with aperture angle greater than $\pi/2$ *Opt. Express* **20** 19206–13

Chapter 5

Scalar approximation for the focusing property of an elliptical mirror

Yuhang Wang, Jian Liu, Tong Wang and Jiubin Tan

5.1 Introduction

In chapters 2–4, we have discussed the imaging characteristics of an elliptical mirror based on the vector diffraction theory. However, in some applied cases, the traditional scalar diffraction theory is still a good approximation for the research on the propagation rule of light. So this chapter will analyze the specific expression of the apodization factor of the elliptical mirror, and establish the focusing property of the scalar diffraction model of the elliptical mirror.

This chapter will study the focusing property of the elliptical mirror based on the scalar diffraction theory, model the scalar diffraction theory-based elliptical mirror focusing in the condition of high numerical aperture and then analyze the focusing property of the elliptical mirror, by using a high numerical aperture Debye diffraction model. The influence factors of the focusing property of the imaging system with high numerical aperture will be analyzed in section 5.2 of this chapter. Because of the influences of apodization factor, polarization state and wave aberration, the optical imaging system with high numerical aperture shall adopt the analysis method different from the traditional lens system. Section 5.3 provides the apodization factor of the elliptical mirror and makes a comparison with those of the lens and the parabolic mirror, on the basis of the apodization factor of the lens with high numerical aperture. Section 5.4 analyzes the focusing properties of the elliptical mirrors with a circular aperture and with a ring-shaped aperture, and makes a comparison with those of the parabolic mirror and the traditional lens.

5.2 Influence factors of focusing property

For the imaging system with a high numerical aperture in microscopy, the paraxial approximation of traditional lens diffraction theory cannot be met. The main reason for this is that the influence of such influence factors as apodization factor, polarization state and wave aberration on the imaging characteristics becomes more notable. The actions of influence factors on the imaging system with high numerical aperture will be analyzed in detail below.

5.2.1 Apodization factor

Now, we discuss the focusing property of a circular thin lens. As shown in figure 5.1, it is assumed that the light field distribution on this thin lens is $P(r)$, which is known as the pupil function of the thin lens, but the light beam is focused to the focal point after passing through the lens, so, ideally, the wavefront behind the lens is a spherical wave W, and the light field distribution on the spherical surface is a function $P(\theta)$ that regards the focusing angle θ as a variable, which is known as the apodization factor herein. It is noted that in the analysis of the lens with a high numerical aperture, the mapping relationship between $P(r)$ and $P(\theta)$ is known as the apodization factor, which is different from the name herein.

It is evident that in the case of low numerical aperture, the light field distribution on the thin lens can be regarded as approximately equal to that of focusing spherical wave, i.e. $P(r) = P(\theta)$. If the numerical aperture is more than 0.75, the difference between $P(r)$ and $P(\theta)$ becomes significant. At this time, the light field distribution must be revised by the apodization factor. The form of apodization factor will depend on many factors, including the refractive index of objective, the design principle of objective, the introduction of spatial filter in front of objective, etc.

5.2.2 Polarization state

In the case of high numerical aperture, the focal location of the linearly polarized light in the objective will no longer maintain its original polarization state. In other words, if the electric field of the incident light is along the axis x, there will be certain electric field strengths in directions y and z when the focal location is the focal point. This will cause the fact that, when the light beam is incident as a linearly polarized light, the focal spot at the focal point no longer has lateral rotational symmetry, as shown in figure 5.2.

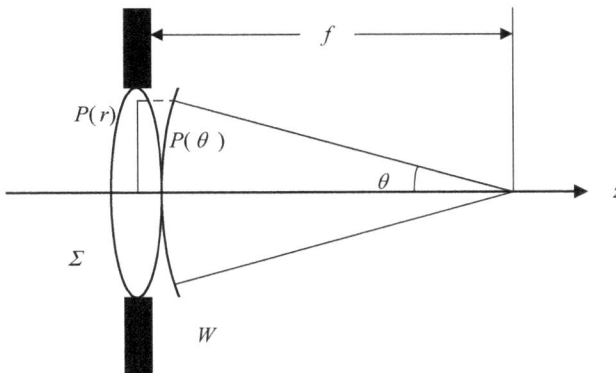

Figure5.1. Lens focusing diagram.

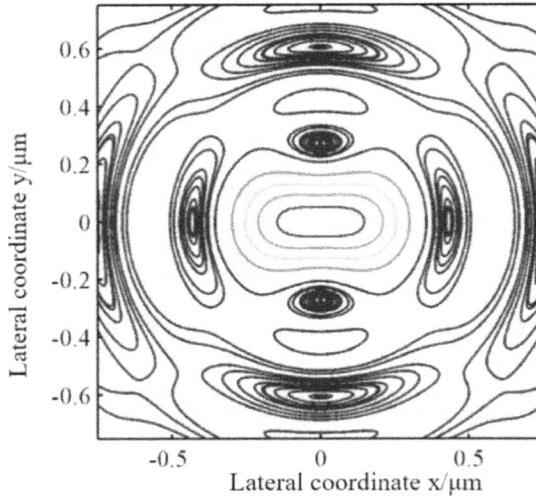

Figure 5.2. Lateral spot of the elliptical mirror (NA = 0.95). (If the polarized light is incident along axis *x*.)

This figure shows the incidence of the linearly polarized light along the axis *x*. In this case, the light field distributions of the focal plane spots of the elliptical mirror with a high numerical aperture in the directions *x* and *y* are no longer in agreement. Such a phenomenon has been detailed and discussed in the analysis on the focused vectors of the relevant objective with a high numerical aperture [1], not to be discussed in this chapter.

5.2.3 Wave aberration

In general, there are limited diffraction and machined surface shape error, assembly and adjustment error and other errors of the lens and mirror in the actual optical system. Accordingly, the actual optical system always has aberration residuals, which more or less deforms the exit wave, making it no longer the ideal spherical wave. The deviation of this deformed actual wave surface with respect to the ideal wave surface is known as wave aberration. In fact, the lens also has aberration, but the reflection type system overcomes this problem, which is one of the potentials of application of the elliptical mirror to the microscopical imaging system.

The function $P(\theta)$ with wave aberration can be written as follows:

$$P(\theta) = P_0(\theta) \exp[-ik\Phi(\theta)] \tag{5.1}$$

where $P_0(\theta)$ and $\Phi(\theta)$ are real functions, respectively representing the amplitude and phase of the light field distribution. $\Phi(\theta)$ is the so-called aberration function. If the numerical aperture is high, the expression form of aberration function is more complex, and the influence on the imaging characteristics is more remarkable.

5.3 Apodization factor of elliptical mirror

This section begins with the apodization factors corresponding to the thin lens with high numerical aperture in different design conditions, and then establishes the apodization factor of the elliptical mirror in the sine condition. By comparing with the apodization factors of the traditional lens and the parabolic mirror, it is found that with the enlarging of the focusing angle, the apodization factors of the elliptical mirror and the parabolic mirror will be increased monotonically, while the apodization factor of the traditional lens will be decreased monotonically. Therefore, for the elliptical mirror and the parabolic mirror, enhancing high-frequency information is favorable for obtaining more detailed information and improving the imaging resolution.

5.3.1 Apodization factor of thin lens

In the optical imaging system, especially in a diffraction-limited system, an important factor that describes the imaging characteristics is pupil function $P(r)$. For the micro lens with high numerical aperture, $P(r)$ is no longer equal to the apodization factor $P(\theta)$, $P(r)$ is the lateral (perpendicular to axis z) light intensity distribution, and $P(\theta)$ is the light intensity distribution on the surface of focusing spherical wave W. The form of apodization factor will depend on many factors, including refractive index of objective, design principle, introduction of spatial filter, etc.

  If the microscope objective meets the sine condition, the two-dimensional space on the focal plane will not be changed. For this reason, most of commercial microscope objectives are designed according to the sine condition, and the analysis on focusing and imaging characteristics of lens mentioned herein are the results from the sine condition. In this chapter, the mapping factor of thin lens with high numerical aperture is defined as follows:

$$g(\theta) = \sin \theta. \tag{5.2}$$

  The definition of apodization factor is shown as follows:

$$P(\theta) = P(r)\sqrt{\cos \theta}. \tag{5.3}$$

5.3.2 Apodization factor of the elliptical mirror with rotational symmetry

The apodization factor of the elliptical mirror will be solved below, as shown in figure 5.3. Similar with the entrance pupil function of the round plane of the lens, the elliptical mirror also has a pupil function and the wavefront distribution of the entrance pupil function is just like W_0 as shown in this figure, which is a divergent spherical wave. After passing through the elliptical mirror, the focusing spherical wavefront distribution is just like W, but the light field distribution on W is expressed still by $P(\theta)$, i.e. spherical apodization factor.

  First, we should get the mapping factor $g(\theta)$ between the object space aperture angle α and the image space aperture angle θ. According to the geometrical imaging

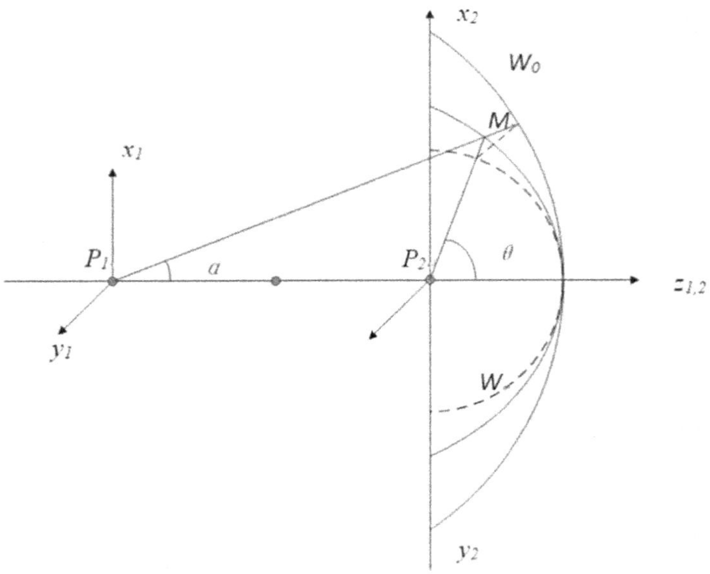

Figure 5.3. Schematic diagram for deduction of apodization factor and imaging characteristic of elliptical mirror.

relationship, by taking point P_1 of the coordinate system shown in figure 5.3 as the origin, the expression of the elliptical mirror is as follows:

$$\frac{r^2}{b^2} + \frac{(z - c)^2}{a^2} = 1 \tag{5.4}$$

where, a and b are respectively the long axis and the minor axis of the elliptical mirror with rotational symmetry, c is the focal distance of the ellipsoid, and $c^2 = a^2 - b^2$. Let the reflection point of the light incident at angle α on the elliptical reflector be $M(r, z)$, where $r^2 = x^2 + y^2$, if $\tan \alpha = r/z$ and $\tan \alpha = r/(z - 2c)$, then

$$\tan \alpha = \frac{(z - 2c)\tan \theta}{z}. \tag{5.5}$$

Based on the geometric property of the elliptical mirror, if and

$$|MP_2| = \frac{z - 2c}{\cos \theta} \tag{5.6}$$

$$|P_1M| = 2a - |MP_2| = 2a - \frac{z - 2c}{\cos \theta} \tag{5.7}$$

for $\triangle P_1MP_2$, according to the cosine law, then

$$|P_1M|^2 = |MP_2|^2 + (2c)^2 - 2|MP_2| \cdot 2c \cdot \cos(\pi - \theta). \tag{5.8}$$

5-5

By substituting formulas (5.6) and (5.7) into (5.8), the result is as follows

$$\left(2a - \frac{z - 2c}{\cos\theta}\right)^2 = \left(\frac{z - 2c}{\cos\theta}\right)^2 + 4c^2 + 4c\cos\theta \cdot \frac{z - 2c}{\cos\theta}. \tag{5.9}$$

Then, the resultant relationship between z and θ is as follows:

$$z = \frac{(a^2 - c^2)\cos\theta}{c \cdot \cos\theta + a} + 2c. \tag{5.10}$$

By substituting formulas (5.10) into (5.5), the result is as follows

$$g(\theta) = \alpha = \arctan\left[\frac{(a^2 - c^2)\sin\theta}{(a^2 + c^2)\cos\theta + 2ac}\right]. \tag{5.11}$$

The apodization factor of the elliptical mirror will be deduced below according to the energy conservation law. Assume that point P_1 is in ideal spot lighting, the light field distribution on the incident wavefront W_0 will be written into $U(\alpha) = P(\alpha)E(\alpha)$, where $P(\alpha)$ is a real function, and a constant in the ideal spot lighting, and $E(\alpha)$ refers to divergent spherical wavefront. Accordingly, the incident light energy corresponding to the surface element is $P(\alpha)^2\delta S_0$, where δS_0 indicates the ring-shaped surface element on W_0 corresponding to da. After the ellipsoid reflection, the amplitude of the reflected light in the wavefront W is $P(\theta)$, a real function, and the relevant reflected light energy is $P(\theta)\delta S$. Accordingly, δS indicates the ring-shaped surface element on W corresponding to $d\theta$. Using the geometrical relationship shown in figure 5.3, we can obtain the equation as follows:

$$\delta S_0 = 2\pi(a + c)^2 \sin\alpha da = 2\pi(a + c)^2 \sin[g(\theta)]g'(\theta)d\theta \tag{5.12}$$

$$\delta S = 2\pi(a - c)^2 \sin\theta d\theta \tag{5.13}$$

where $g'(\theta)$ indicates the first derivative of $g(\theta)$ that depends on the variable θ. According to the energy conservation law,

$$P(\alpha)^2\delta S_0 = P(\theta)^2\delta S. \tag{5.14}$$

By substituting formulas (5.12) and (5.13) into the above equation, we can get the final apodization factor of the elliptical mirror, as follows:

$$P(\theta) = P(\alpha)\left|\frac{a + c}{a - c} \times \frac{\sin[g(\theta)]}{\sin\theta}g'(\theta)\right|. \tag{5.15}$$

The formula (5.15) is meaningful to the elliptical mirrors with different geometric parameters. Figure 5.4 shows the comparison of the apodization factors of the elliptical mirror, parabolic mirror, and traditional lens.

The apodization factor of the traditional lens is as follows

$$P_{lens}(\theta) = \sqrt{\cos\theta}. \tag{5.16}$$

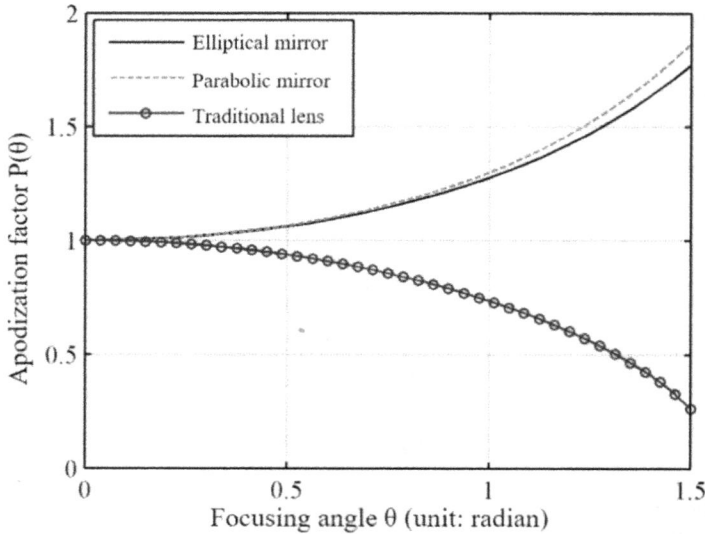

Figure 5.4. Apodization factor curve of different optical elements (c/a = 3/5 for elliptical mirror).

The apodization factor of the parabolic mirror is as follows

$$P_{parabolic}(\theta) = \frac{2}{1 + \cos\theta}. \tag{5.17}$$

In figure 5.4, we can observe that with the enlarging of the focusing angle, the apodization factors of the elliptical mirror and the parabolic mirror will be increased monotonically, while the apodization factor of the traditional lens will be decreased monotonically. If the radian of the focusing angle is 1.5 (around $\pi/2$), the values of the apodization factors of these two mirrors approach 2, while that value of the traditional lens is 0.25. This means that for the elliptical mirror and the parabolic mirror, enhancing high-frequency information is favorable for obtaining more detailed information and improving the imaging resolution. Particularly, as shown in figure 5.4, the curve of the apodization factor of the parabolic mirror is much steeper than that of the elliptical mirror. But it is necessary to point out that if the half focal distance c of the elliptical mirror takes different values, there will be a difference in its curve. The closer to the semi-major axis a of the ellipsoid, the steeper the curve is, as shown in figure 5.5. But the processing cost and difficulty of the elliptical mirror are increased simultaneously.

5.4 Analysis on focusing property of elliptical mirror

The focusing property of the optical system, i.e. the light field distribution near the focal point of the optical system, is the most intuitive criterion of the system resolution. This section will analyze the focusing property of the elliptical mirror in the ideal spot lighting condition. Here, we use the full width at half maximum (FWHM) of the main lobe of the response curve to provide the quantitative

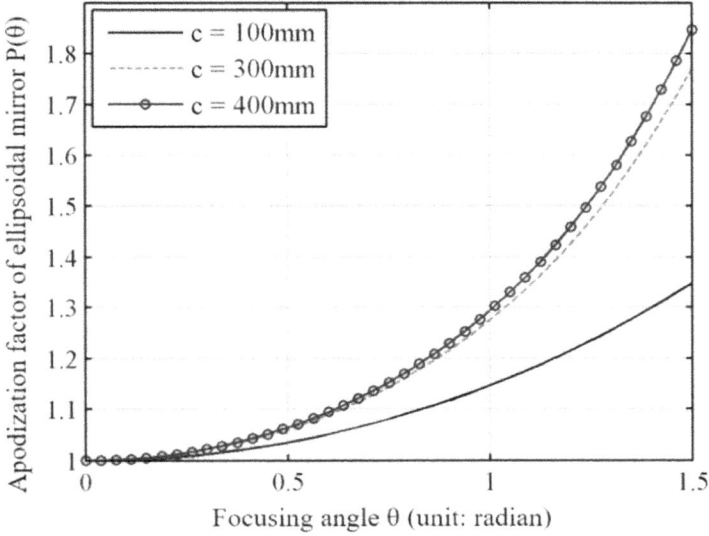

Figure 5.5. Apodization factor curve of elliptical mirror with different geometric parameters ($a = 500$ mm).

description of the lateral and axial focusing spots. Because the actual optical system is in circular symmetry, we must first analyze the focusing property of the elliptical mirror with circular aperture, and then analyze that of the elliptical mirror with a ring-shaped aperture and compare both, in conjunction with the practical problems in application of the elliptical mirror.

5.4.1 Focusing property of elliptical mirror with circular aperture

Based on the apodization factor given in the above section, we can further obtain the focusing property of the scalar diffraction model of the elliptical mirror. As shown in figure 5.3, under the condition of Debye approximation, $P(\theta)$ indicates the apodization factor of the elliptical mirror, and if P_1 is in the spot lighting, the light field distribution at the focal point P_2 of the elliptical mirror will be expressed as

$$U(v, u) = \frac{2\pi i}{\lambda} \exp(-ikx) \int_0^\gamma P(\theta) J_0\left(\frac{v \sin \theta}{\sin \gamma}\right) \exp\left(\frac{iu \sin^2(\theta/2)}{2 \sin^2(\gamma/2)}\right) \sin \theta d\theta \quad (5.18)$$

where γ indicates the maximum focusing angle of the elliptical mirror, the limit of integration for the elliptical mirror with circular aperture is $[0, \gamma]$, and the expression of $P(\theta)$ is as shown in the formula (5.15). v and u indicate lateral and axial optical coordinates respectively:

$$v = \frac{2\pi}{\lambda} r \sin \gamma \quad (5.19)$$

$$u = \frac{8\pi}{\lambda} z \sin^2\left(\frac{\gamma}{2}\right). \quad (5.20)$$

The focusing property of the elliptical mirror, i.e. the light intensity distribution $I(u,v)$ at the focal point P_2, is expressed as

$$I(v, u) = \frac{4\pi^2}{\lambda^2} \left| \int_0^\gamma P(\theta) J_0\left(\frac{v \sin \theta}{\sin \gamma}\right) \exp\left(\frac{iu \sin^2(\theta/2)}{2 \sin^2(\gamma/2)}\right) \sin \theta d\theta \right|^2. \qquad (5.21)$$

Figure 5.6 shows the three-dimensional light intensity distribution on the meridian plane of focusing spot of the elliptical mirror, and the relevant geometric parameters of the elliptical mirror are as follows: $a = 500$ mm, and $c = 300$ mm. Under the condition of Debye approximation, the difference of focusing property of the thin lens, the parabolic mirror and the elliptical mirror is mainly reflected in the apodization factor. So, the formula (5.16) is the common expression for the three focusing properties, but because the apodization factor of the elliptical mirror is complex in form, compared with the thin lens and the parabolic mirror, we typically make direct calculation without giving the specific expressions.

Accordingly, the lateral response characteristic on the focal plane of the elliptical mirror is as follows:

$$I(v, u = 0) = -\frac{4\pi^2}{\lambda^2} \left| \int_0^\gamma P(\theta) J_0\left(\frac{v \sin \theta}{\sin \gamma}\right) \sin \theta d\theta \right|^2. \qquad (5.22)$$

The axial response characteristic on its meridian plane is as follows:

$$I(v = 0, u) = \frac{4\pi^2}{\lambda^2} \left| \int_0^\gamma P(\theta) \exp\left(\frac{iu \sin^2(\theta/2)}{2 \sin^2(\gamma/2)}\right) \sin \theta d\theta \right|^2. \qquad (5.23)$$

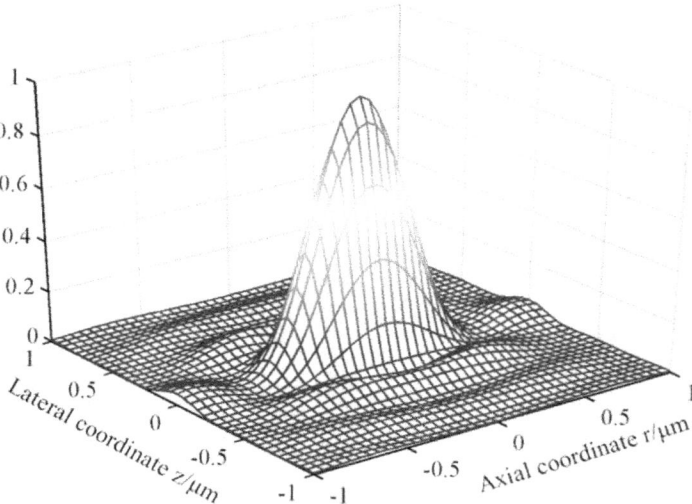

Figure 5.6. Three-dimensional representation of the focal spot of the elliptical mirror on a meridian plane.

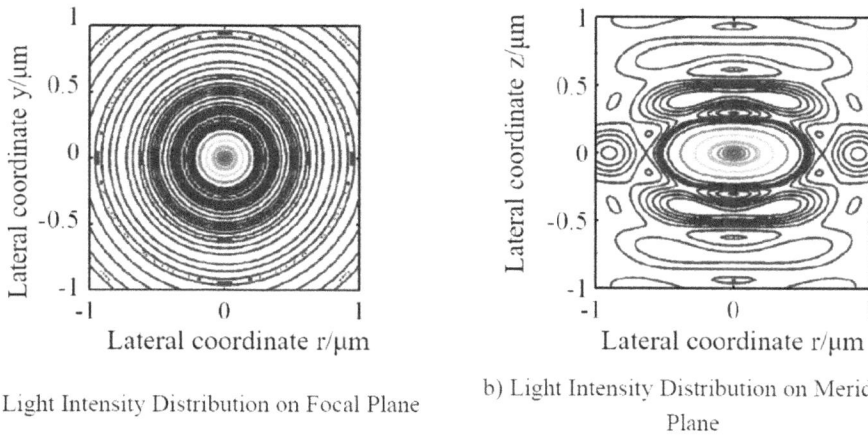

a) Light Intensity Distribution on Focal Plane

b) Light Intensity Distribution on Meridian Plane

Figure 5.7. Light intensity distribution near the focal point of the elliptical mirror with circular aperture.

Figures 5.7(a) and (b) show the light intensity distributions of the focal plane and the meridian plane respectively. As shown, the focal spot is a three-dimensional ellipsoid.

As shown in figure 5.8, the elliptical mirror has the following geometric parameters: $a = 500$ mm, and $c = 300$ mm. Figures 5.8(a) and (c) respectively are the lateral and axial focusing property curves of the elliptical mirror, traditional lens and parabolic mirror if the maximum focusing angle is $\pi/3$. The corresponding objective numerical aperture is 0.85. In these two figures, we can observe that their axial response curves coincide with each other roughly, and the lateral FWHM of the elliptical mirror and parabolic mirror is slightly narrower than that of the lens. For this reason, it is believed that these three systems have the approximately same resolution when imaging. Figures 5.8(b) and (d) show that if the maximum focusing angle is $\pi/2$ or the objective numerical aperture is 1.0, their FWHMs are increased accordingly; compared with the numerical aperture of 0.85, both the lateral resolution and the axial resolution are increased; and the side lobe of the lateral spot is enhanced and the side lobe of the axial spot is unchanged basically, which is coherent with the traditional lens. In these figures, we can clearly observe that the lateral and axial FWHMs of the elliptical mirror and the parabolic mirror are 90% of the lens. It can be concluded that the objective system, consisted of the elliptical mirror with high numerical aperture and of the parabolic mirror with high numerical aperture, has better image resolution than the traditional lens system.

In particular, the above mentioned lens is the perfect lens. In fact, the objective (non-oil immersion dry objective) made by the lens is always restricted in the numerical aperture less than 1. During the comparison between the elliptical mirror and the parabolic mirror, we can observe that although they can make the numerical aperture more than or equal to 1, the elliptical mirror system adopts point illumination which is in favor of aberrational correction of incident wavefront, and there is no difficulty in incident beam shaping for the parabolic mirror with a high numerical aperture.

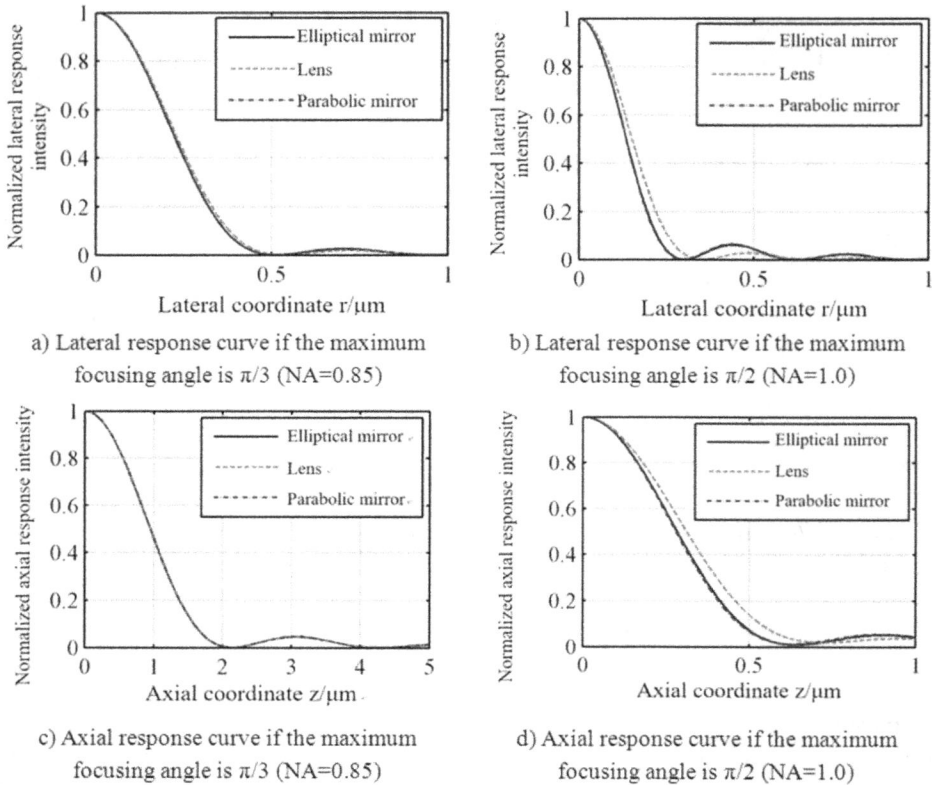

a) Lateral response curve if the maximum
focusing angle is $\pi/3$ (NA=0.85)

b) Lateral response curve if the maximum
focusing angle is $\pi/2$ (NA=1.0)

c) Axial response curve if the maximum
focusing angle is $\pi/3$ (NA=0.85)

d) Axial response curve if the maximum
focusing angle is $\pi/2$ (NA=1.0)

Figure 5.8. Comparison of focusing properties of different optical elements.

Figure 5.9 shows the comparison of focusing properties of the elliptical mirrors with different numerical apertures ($a = 500$ mm, $c = 300$ mm). In this figure, we can observe that with the growth of the numerical aperture, the FWHMs of the lateral and axial response curves are narrowed accordingly; if the numerical aperture is 1.0, the FWHM of the lateral response curve of the elliptical mirror is only 40% for 0.7; and the axial resolution is increased significantly and FWHM is only 14.3% for 0.7. Meanwhile, the side lobe of the lateral response curve is enhanced while the side lobe of the axial response curve is essentially constant. This phenomenon is also in agreement with the conclusion gained in the analysis of the lens system with a high numerical aperture.

5.4.2 Focusing property of elliptical mirror with ring-shaped aperture

In the practical application of the elliptical mirror, there may be a central obscuration caused by an objective table or detection system (which is described by an obscured focusing angle and is known as obscured numerical aperture). At this point, the elliptical mirror has a ring-shaped aperture and its focusing angle range is

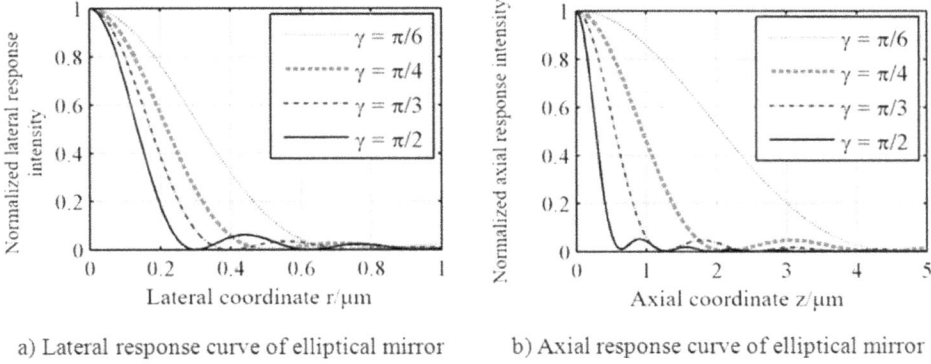

a) Lateral response curve of elliptical mirror   b) Axial response curve of elliptical mirror

Figure 5.9. Light intensity distribution of elliptical mirror with different maximum focusing angles (NA).

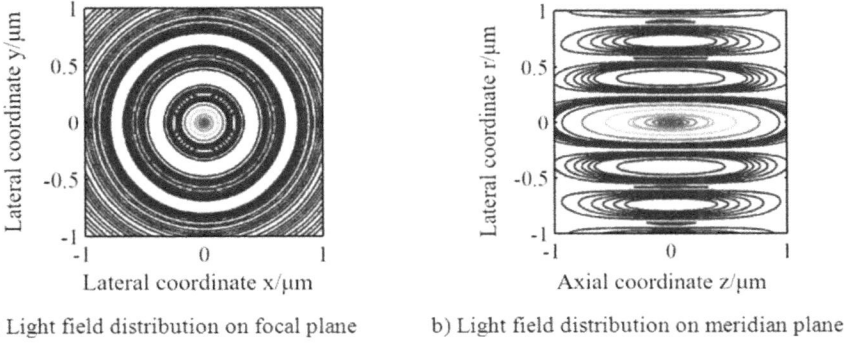

a) Light field distribution on focal plane   b) Light field distribution on meridian plane

Figure 5.10. Light field distribution nearby the focal point of the elliptical mirror with a ring-shaped aperture and obscured focusing angle $\varepsilon = \pi/3$.

changed to $[\varepsilon, \gamma]$ $(0 < \varepsilon < \gamma)$. But in the expression of the focusing property, only the limit of integration is changed, as follows:

$$I_{annular}(v, u) = \frac{4\pi^2}{\lambda^2} \left| \int_0^\gamma P(\theta) J_0\left(\frac{v \sin \theta}{\sin \gamma}\right) \exp\left(\frac{iu \sin^2(\theta/2)}{2 \sin^2(\gamma/2)}\right) \sin \theta d\theta \right|^2 . \quad (5.24)$$

Figure 5.10 shows the simulation diagram of light intensity distribution on focal plane and meridian plane if the obscured focusing angle is $\pi/3$. By comparing with the circular aperture spots shown in figure 5.10(a) and in figure 5.7(a), we can clearly observe that the FWHM of the main lobe is increased, which means that the ring-shaped elliptical mirror has higher lateral resolution. From the frequency domain, we can say that the ring-shaped pupil is equivalent to a wave filter with annular space, and we can consider that the central obscuration will stop the propagation of low-frequency information and enhance the high-frequency detailed information, thereby increasing the resolution, but causing the side lobe energy to be enhanced for reducing the signal-to-noise ratio. In figure 5.10(b), we can observe that, compared with the circular aperture, the focal depth of the ring-shaped aperture is increased

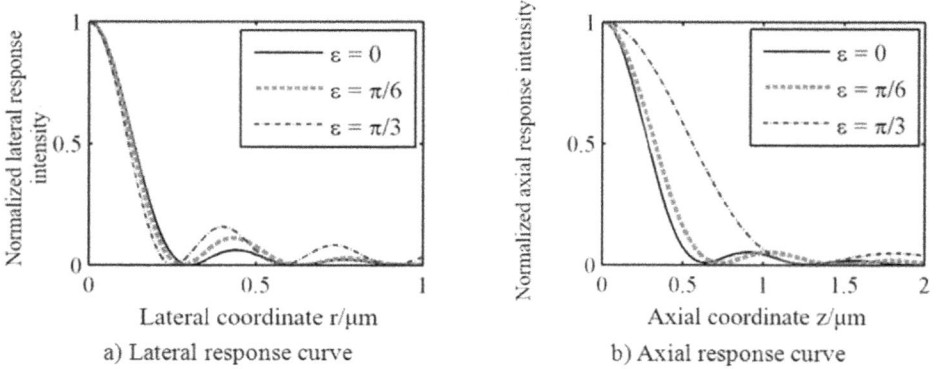

a) Lateral response curve   b) Axial response curve

Figure 5.11. Comparison of focusing properties of elliptical mirrors with different obscured focusing angles.

notably, and the axial resolution is reduced accordingly. Figure 5.11 shows the comparison of focusing properties corresponding to different obscured focusing angles. The above analysis on the ring-shaped elliptical mirror is consistent with the traditional ring-shaped lens.

5.5 Comparative analysis on vector diffraction model

The scalar diffraction theory model proposed in this article is an approximation of the vector diffraction theory model. In order to verify that the scalar diffraction theory model of the focusing property of elliptical mirror is valid, we will compare, simulate and analyze this model and the vector diffraction theory model proposed in chapter 4 and reference [2].

In [2], it is pointed out that the three-dimensional vector function expression of the focusing field of the elliptical mirror with a high numerical aperture is as follows:

$$\begin{cases} e_s^x = iA\left(I_0 + I_2 \cos\left(2\phi_s\right)\right) \\ e_s^y = iAI_2 \sin\left(2\phi_s\right) \\ e_s^z = -2AI_1 \cos\phi_s \end{cases} \tag{5.25}$$

where,

$$A = kf/2\sqrt{2} \tag{5.26}$$

$$I_0 = \int_0^{\theta_{max}} P(\theta)\sin\theta(1 + \cos\theta)J_0(k\rho_s \sin\theta_s \sin\theta)e^{-ik\rho_s \cos\theta_s \cos\theta}d\theta \tag{5.27}$$

$$I_1 = \int_0^{\theta_{max}} P(\theta)\sin^2\theta J_1(k\rho_s \sin\theta_s \sin\theta)e^{-ik\rho_s \cos\theta_s \cos\theta}d\theta \tag{5.28}$$

$$I_2 = \int_0^{\theta_{max}} P(\theta)\sin\theta(1 - \cos\theta)J_2(k\rho_s \sin\theta_s \sin\theta)e^{-ik\rho_s \cos\theta_s \cos\theta}d\theta. \tag{5.29}$$

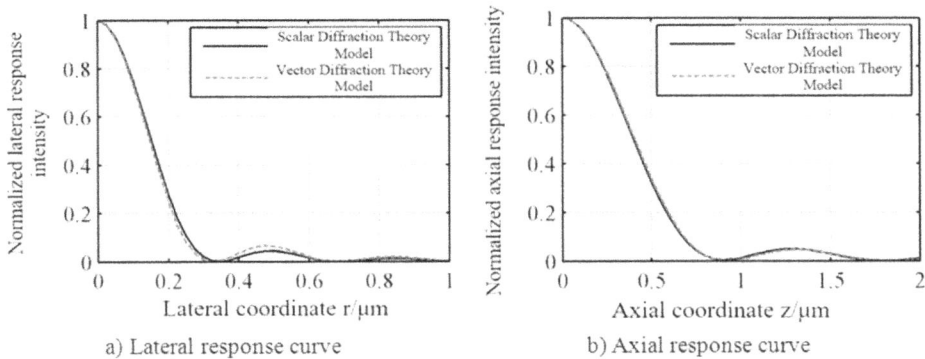

Figure 5.12. Comparison between scalar diffraction theory model and vector diffraction theory model of focusing property of elliptical mirror.

$P(\theta)$ of the above expression refers to the apodization factor of the elliptical mirror. The response curve comparison between this vector diffraction theory model and the scalar diffraction theory model proposed in this article can be obtained through numerical analysis, as shown in figure 5.12.

The lateral response curve of vector theory is the response curve of focal plane in y-axis direction if the linear polarized light is incident in direction x. In this figure, we can observe that the lateral and axial response curves are essentially coincided, which means that the scalar diffraction theory model proposed herein has the same result as the vector diffraction theory model proposed in the references. It can also serve as evidence that the theory model proposed herein is valid.

5.6 Summary

This chapter gives the specific expression of the apodization factor of elliptical mirror, establishes the focusing property of the scalar diffraction model of elliptical mirror and draws a conclusion that we can overcome the problem that the traditional lens has high lateral and axial resolutions and a low numerical aperture less than 1 by means of the elliptical mirror, by comparing the focusing properties of the parabolic mirror and the traditional lens. Simulation results show that the lateral and axial FWHMs of the elliptical mirror and the parabolic mirror are 90% of the lens and the side lobe is slightly enhanced if the numerical aperture (NA) is equal to 1. However, compared with the parabolic mirror, the elliptical mirror system adopts spot lighting and has the advantage of simple aberrational correction.

References

[1] Gu M 2000 *Advanced Optical Imaging Theory* (Berlin: Springer) pp 71–109 143–198
[2] Liu J, Tan J B and Wilson T 2012 Rigorous theory on elliptical mirror focusing for point scanning microscopy *Opt. Express* **20** 6175–84

IOP Publishing

Elliptical Mirrors
Applications in microscopy
Jian Liu

Chapter 6

Aberration analysis of an elliptical mirror with a high numerical aperture

Chenguang Liu, Tong Wang, Jian Liu and Jiubin Tan

6.1 Introduction

Compared with the traditional lens systems, the reflective imaging system has the most important advantage that it does not need chromatic aberration correction or compensation, but it has higher requirements for surface shape accuracy, which will cause an increase in processing difficulty and cost. In practical use, various aberrations will be caused by processing errors, installation and adjustment errors, and imperfection of optical elements, and the aberration with great influence is the primary aberration, also called the Seidel aberration. The correlation analysis indicates that the elliptical mirror is sensitive to aberrations, so the analysis of the aberration characteristics of the elliptical mirror is an important issue in the elliptical reflective confocal imaging theory.

The traditional aberration analysis method is mainly to research the influence of various aberrations on image on the basis of geometrical optics. An image is regarded as a figure formed by the intersection of geometrical ray and image plane but becomes obscure due to loss of acuteness during ray focusing. The geometrical optics provides an approximate model which is effective only in the extreme case of large aberration or short wavelength, so it can be expected that when the aberration decreases (the wave aberration is the order of wavelength or smaller), the geometrical optics will gradually lose its effectiveness, and then diffraction plays an important role. At this moment, the geometry theory of the aberration must be supplemented with the more strict diffraction theory [1].

This chapter studies the aberration characteristics of the elliptical mirror on the basis of the scalar diffraction model of the elliptical mirror with a high numerical aperture, establishes a diffraction integral model of the elliptical mirror with a high

doi:10.1088/978-0-7503-1629-3ch6

numerical aperture in the presence of the aberration, and analyzes the influence of the primary aberration on the focusing characteristics.

The research of the geometrical aberration can directly guide the processing, installation and commissioning of the relevant optical elements, and the geometry theoretical analysis of the aberration of the elliptical mirror is conducted in section 6.2 to provide the primary aberration coefficient of the elliptical mirror through the third order approximation of the reflected ray equation. In section 6.3, a diffraction model of the aberration of an optical system with a high numerical aperture is established as the theoretical basis of aberration analysis. To independently analyze the aberrations of different forms, the aberration function shall be expanded, so Zernike circle polynomial expansion is conducted on the aberration in section 6.4, thereby obtaining the common expression of the aberration function. In section 6.5, simulation analysis is respectively conducted on the influences of the primary spherical aberration, primary coma, primary astigmatism, field curvature and distortion on the focusing characteristics of the elliptical mirror, and quantitative analysis is conducted on the tolerance for aberration condition of the elliptical mirror for each primary aberration.

6.2 Analysis of geometrical aberration of elliptical mirror

The geometrical aberration is an approximate model which has important guiding significance on the requirement for processing precision and the design of the installation and commissioning system of the elliptical mirror, so the geometrical aberration and its coefficient are analyzed before the influence of the aberration on the focusing characteristics is discussed on the basis of the diffraction theory.

6.2.1 Reflected ray formula of elliptical mirror

The geometry expression of an ideal elliptical mirror can be written as

$$h^2 = 2rz - (1 + k)z^2 \quad (-1 < k < 0) \tag{6.1}$$

where $h = \sqrt{x^2 + y^2}$, and k represents the surface shape coefficient of a quadric surface.

$$k = -\left(\frac{s - s'}{s + s'}\right)^2. \tag{6.2}$$

In a practical application, the explicit expression in the form of $z = f(h)$ is frequently used, and (6.1) can be expressed as

$$z = \frac{h^2/r}{1 + \sqrt{1 - (1 + k)h^2/r^2}}. \tag{6.3}$$

To obtain the geometrical aberration coefficient of the elliptical mirror, we need to calculate the expression of the reflected ray of the incident ray. The expression of

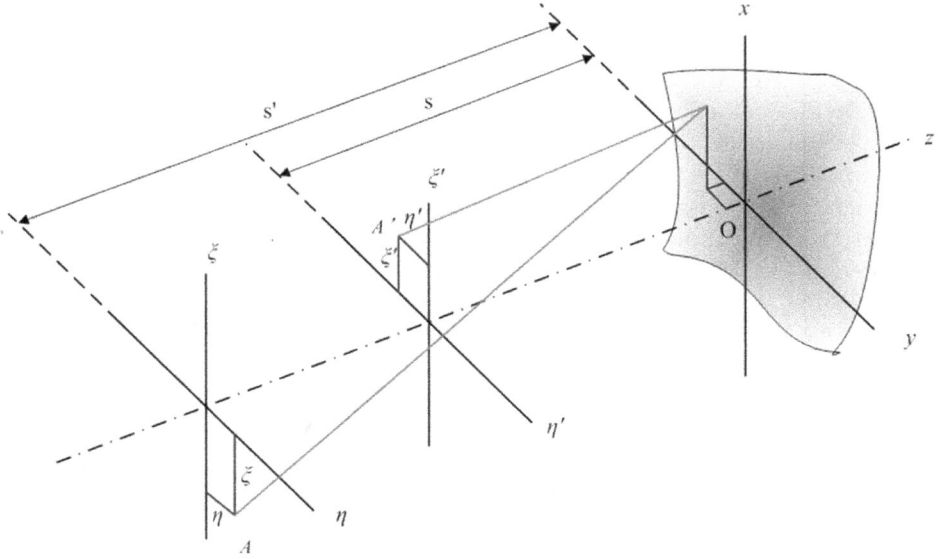

Figure 6.1. The geometrical relationship of ray reflection.

the given incident ray and its reflected ray on the elliptical reflective plane in the Cartesian coordinate system can be obtained according to the theory in the literature [2] by D Korsch. Ray tracing is shown in figure 6.1.

Rays are emitted from $A(\xi, \eta)$ on the object plane and reflected at the point $P(x, y, z)$ on the reflective plane, and the reflected rays intersect with the image plane at $A'(\xi', \eta')$. The object plane and the image plane are perpendicular to the optical axis x, and their distances from the coordinate origin of the reflective plane are respectively s and s'.

The first-order partial derivative of $z = f(x, y)$ with respect to x and y is expressed by z_x and z_y. Then the calculation formula of the reflected ray of the elliptical mirror is

$$\begin{cases} \xi' = (Ux + V\xi)/W \\ \eta' = (Uy + V\eta)/W \end{cases} \tag{6.4}$$

where,

$$U = r[r(s + s') - 2ss'] + 2[r^2 - (rs + rs' - ss)(1 + k)]z$$

$$- [2r - (s + s')(1 + k)]kz^2 + 2(s' - r + kz)(x\xi + y\eta)$$

$$V = -r^2s' + r[r + 2s'k]z - [2r + (1 + k)s']kz^2 + k(1 + k)z^3 \tag{6.5}$$

$$W = r^2s - 2[r - (1 + k)z](x\xi + y\eta) + r[3r - 2s(2 + k)]z$$

$$- [2r(1 + 2k) - s(1 + k)(2 + k)]z^2 + k(1 + k)z^3.$$

It is easy to know that the rotationally symmetric elliptical mirror has the characteristic of aplanatic image formation. The aplanatic point means that incident rays emitted from any point can form an ideal image at another point with the geometrical aberration precision regardless of the incident aperture angle, and this imaging relationship is called aplanatic image formation [3].

For the parabolic mirror ($k = -1$), this pair of aplanatic points are a finite point and an infinite point, wherein the distance of the finite point is $r/2$, which is called the focal point of the paraboloid. For the elliptical mirror ($-1 < k < 0$), this pair of aplanatic points are at the same side of the vertex of the reflective plane and are used as the two focuses of the ellipsoid. The elliptical reflective plane has a pair of finite separate aplanatic points, so the elliptical mirror can be in the form of point illumination, and does not have requirements for broad beam collimation as strict as the parabolic mirror when the entrance pupil diameter is large.

Under the paraxial approximation conditions, the optical system can be regarded as ideal Gaussian imaging. But obviously, there is no optical system for ideal imaging, and especially, the reflective system has more complex imaging characteristics than the lens system, so it is necessary to discuss the degree of deviation of the actual rays from the Gaussian optics system—aberration. Within the framework of geometrical optics, especially for Seidel theory precision, terms with the off-axis distance more than quadratic in the eigenfunction expansion shall be reserved during the analysis of the expected deviation of the ray path from the Gaussian theory. These terms represent geometrical aberrations, and we only study the aberration with the lowest level, which is usually known as primary aberration or Seidel aberration. Therefore, it is necessary to conduct third-order approximation on the reflected ray formula of the elliptical mirror.

Before the series expansion of the reflected ray formula of a single elliptical mirror, series expansion shall be firstly conducted on the surface shape equation (6.2) of the mirror as follows

$$z = \frac{1}{2r}h^2 + \frac{1+k}{8r^3}h^4 + \cdots. \tag{6.6}$$

In the following calculation, the first term of z is only needed, and then the above formula is approximated as

$$z = \frac{h^2}{2r}. \tag{6.7}$$

Obviously, (6.4) is a parabolic equation. In view of imaging on a pair of conjugate planes, to discuss the distribution of the aberration on the axis, an axial deviation $\Delta s'$ of the image plane is introduced, and s' is replaced with $s' + \Delta s'$ in the formula. Without loss of generality, supposing $\Delta s'$ is a high-order infinitesimal, and $o(\Delta s') = 0$, formula (6.4) is substituted into formula (6.3), terms of higher order are ignored, and then the intersection of the reflected ray and the image plane can be expressed as

$$\xi' = \frac{(m-1)^3}{8s'^2}\Delta k \cdot h^2 x + \frac{m^2-1}{4ss'}[2(x\xi + y\eta)x + h^2\xi]$$
$$+ \frac{m-1}{s^2}(x\xi + y\eta)\xi + m\left(1 + \frac{\Delta s'}{s'}\right)\xi - \frac{\Delta s'}{s'}x$$
$$\eta' = \frac{(m-1)^3}{8s'^2}\Delta k \cdot h^2 y + \frac{m^2-1}{4ss'}[2(x\xi + y\eta)y + h^2\eta]$$
$$+ \frac{m-1}{s^2}(x\xi + y\eta)\eta + m\left(1 + \frac{\Delta s'}{s'}\right)\eta - \frac{\Delta s'}{s'}y$$

$$(6.8)$$

where, Δk is the surface shape function of the actual elliptical mirror, and different from the surface shape coefficient of the corresponding aplanatic image formation surface, it is expressed as

$$\Delta k = k - k_0 = k + \left(\frac{s-s'}{s+s'}\right)^2. \tag{6.9}$$

Finally, the reflected ray formula of the elliptical mirror in third-order approximation expressed by Δk and lateral magnification $m = -s'/s$ of the optical system is obtained:

$$\xi' = \frac{(m-1)^3}{8s'^2}\Delta k \cdot h^2 x + \frac{m^2-1}{4ss'}[2(x\xi + y\eta)x + h^2\xi]$$
$$+ \frac{m-1}{s^2}(x\xi + y\eta)\xi + m\left(1 + \frac{\Delta s'}{s'}\right)\xi - \frac{\Delta s'}{s'}x$$
$$\eta' = \frac{(m-1)^3}{8s'^2}\Delta k \cdot h^2 y + \frac{m^2-1}{4ss'}[2(x\xi + y\eta)y + h^2\eta]$$
$$+ \frac{m-1}{s^2}(x\xi + y\eta)\eta + m\left(1 + \frac{\Delta s'}{s'}\right)\eta - \frac{\Delta s'}{s'}y.$$

$$(6.10)$$

6.2.2 Analysis of the aberration coefficient of a single rotating elliptical mirror

The third-order approximation of the reflected ray formula given in formula (6.7) is an expression with the space coordinates of the point object and the surface shape function of the elliptical mirror as parameters. A pupil function will be introduced in the following discussion. The space coordinates x, y and h of the point object on the object plane are replaced with the coordinates $\tilde{x}$, $\tilde{y}$ and $\tilde{h}$ of the focal point of the ray and the entrance pupil plane, and the transformation relation is

$$x = \frac{1}{s-t}(s\tilde{x}-t\xi)$$
$$y = \frac{1}{s-t}(s\tilde{y}-t\eta). \tag{6.11}$$

Accordingly, $h^2 = x^2 + y^2$, $\tilde{h}^2 = \tilde{x}^2 + \tilde{y}^2$ and $\rho^2 = \xi^2 + \eta^2$ are substituted into formula (6.7), and this formula can be written as an expression with $\tilde{x}$, $\tilde{y}$ and $\tilde{h}$ as parameters and can be rearranged as:

$$
\begin{aligned}
\xi' = {}& \frac{s'}{8}\left[\frac{s(m-1)}{s'(s-t)}\right]^3 \Delta k \cdot \tilde{h}^2 \tilde{x} \\
&+ \frac{m}{8}\left\{t\left[\frac{(m-1)s}{(s-t)s'}\right]^3 \Delta k - 2\frac{m+1}{m-1}\left[\frac{s(m-1)}{s'(s-t)}\right]^2\right\}[\tilde{h}^2\xi + 2(\tilde{x}\xi + \tilde{y}\eta)\tilde{x}] \\
&+ \frac{m^2}{4s'}\left\{t^2\left[\frac{s(m-1)}{s'(s-t)}\right]^3 \Delta k - 4t\frac{m+1}{m-1}\left[\frac{s(m-1)}{s'(s-t)}\right]^2 + 4\left[\frac{s(m-1)}{s'(s-t)}\right]\right\} \\
&\times (\tilde{x}\xi + \tilde{y}\eta)\xi \\
&+ \frac{m^2}{8s'}\left\{t^2\left[\frac{s(m-1)}{s'(s-t)}\right]^3 \Delta k - 4t\frac{m+1}{(m-1)}\left[\frac{s(m-1)}{s'(s-t)}\right]^2\right\}\tilde{x}\rho^2 \\
&+ \frac{m^3}{8s'^2}\left\{t^3\left[\frac{s(m-1)}{s'(s-t)}\right]^3 \Delta k - 6t^2\frac{m+1}{m-1}\left[\frac{s(m-1)}{s'(s-t)}\right]^2 + 8t\left[\frac{s(m-1)}{s'(s-t)}\right]\right\}\rho^2\xi \\
&- \frac{s\Delta s'}{s'(s-t)}\tilde{x} + \frac{t\Delta s'}{s'(s-t)}\xi - \xi\frac{\Delta s'}{s} + m\xi.
\end{aligned}
\tag{6.12}
$$

In a similar way, the expression of η' can be obtained, which will not be repeated here.

The above formula is more complex than formula (6.18). But obviously, this formula can be greatly simplified with the relevant optical parameter expression, and most of the variables are defined in Gaussian optics:

(1) Surface shape curvature of mirror:

$$
c = \frac{1}{r} = \frac{1-m}{2s'}.
\tag{6.13}
$$

(2) Surface shape eccentricity of mirror

$$
e = \frac{1+m}{1-m}.
\tag{6.14}
$$

(3) Image scale factor:

$$
f^* = m(s-t).
\tag{6.15}
$$

(4) Image scale coefficient:

$$
\tau = \frac{s-t}{s}.
\tag{6.16}
$$

(5) Angular magnification at pupil:

$$\mu = \frac{t}{t'} = \frac{t(s + s')}{ss'} - 1. \tag{6.17}$$

Then, the coordinate of the reflected ray of the single quadric surface mirror with entrance pupil function within the image space $s' \pm \Delta s'$ is expressed as

$$\xi' = A \cdot f^* \tilde{h}^2 \tilde{x} - B \cdot m \, [\tilde{h}^2 \xi + 2(\tilde{x}\xi + \tilde{y}\eta)\tilde{x}] + C \cdot \frac{m^2}{f^*}(\tilde{x}\xi + \tilde{y}\eta)\xi$$

$$+ D \cdot \frac{m^2}{f^*} \tilde{x}\rho^2 - E \cdot \frac{m^3}{f^{*2}}\rho^2 \xi + F \cdot f^* \tilde{x}\Delta s' + \frac{1}{f^*}(f^* + \mu \Delta s')m\xi$$

$$\eta' = A \cdot f^* \tilde{h}^2 \tilde{y} - B \cdot m \, [\tilde{h}^2 \eta + 2(\tilde{x}\xi + \tilde{y}\eta)\tilde{y}] + C \cdot \frac{m^2}{f^*}(\tilde{x}\xi + \tilde{y}\eta)\eta \tag{6.18}$$

$$+ D \cdot \frac{m^2}{f^*} \tilde{y}\rho^2 - E \cdot \frac{m^3}{f^{*2}}\rho^2 \eta + F \cdot f^* \tilde{y}\Delta s' + \frac{1}{f^*}(f^* + \mu \Delta s')m\eta$$

wherein, A, B, C, D, E and F are constants and only concerned with the relation between the surface shape function of the mirror and the location of the selected conjugate point, and the specific expression is as follows:

$$A = \tau^{-4}c^3 \Delta k$$

$$B = t\tau^{-3}c^3 \Delta k - \tau^{-2}ec^2$$

$$C = 2(t^2 \tau^{-2}c^3 \Delta k - 2t\tau^{-1}ec^2 + c) \tag{6.19}$$

$$D = t^2 \tau^{-2}c^3 \Delta k - 2t\tau^{-1}ec^2$$

$$E = t^3 \tau^{-1}c^3 \Delta k - 3t^2 ec^2 + 2t\tau c$$

$$F = 1/f^{*2}.$$

On the Gaussian image plane, i.e. $\Delta s' = 0$, the remaining five terms in the aberration formula are called primary aberrations or Seidel aberrations. Parameters A to E are aberration coefficients, and the geometrical aberrations corresponding to all the terms are respectively spherical aberration, coma, astigmatism, field curvature and distortion [2]. For the confocal scanning system, the aberration will greatly influence the light field distribution on the focal plane. When the aberration is large, even it is impossible to focus, the spherical aberration has the greatest influence, and then coma and astigmatism. It can be seen from the following analysis that field curvature and distortion have no influence on the distribution of diffraction light spots.

For a pair of given conjugate surfaces, the aberration coefficient mainly depends on the surface shape coefficient k, the curvature c of the vertex and the location of the pupil of the mirror. Among the five coefficients given in formula (6.10) to formula (6.15), only spherical aberration, coma and distortion are independent, while astigmatism and field curvature are usually studied together. Supposing the field curvature coefficient c_0 is the linear addition of coefficient C and D:

$$c_0 = -(C + 2D). \tag{6.20}$$

The object distance of the elliptical mirror is finite, and C and c_0 are usually studied. The relationships among all aberration coefficients can be summarized below:

$$
\begin{aligned}
A &= \tau^{-4}c^3\Delta k \\
B &= \tau t A + \frac{c}{2\tau^2}(v - {'v}) \\
C &= -2\tau t(\tau t A - 2B) + 2c \\
D &= -\tau t(\tau t A - 2B) \\
E &= -\tau t(\tau t B - C) \\
c_0 &= -2(C - c)
\end{aligned}
\tag{6.21}
$$

where, v and v' are respectively the reciprocals of the object distance and the image distance s and s'.

It can be seen that the relationships among the aberration coefficients of a single quadric surface mirror are as follows:

1. When the curvature c of the vertex is 0, all the aberrations will disappear, and the corresponding reflective plane is a plane mirror, which conforms to the characteristics of ideal imaging of the mirror. This article mainly discusses the elliptical mirror, so supposing $c \neq 0$:
2. When $\Delta k = 0$, $A = 0$; especially, when $\Delta k = 0$, the mirror provides aplanatic image for the pair of given conjugate surfaces. At this moment, spherical aberration can be fully eliminated.
3. If the pupil is on the plane ($t = 0$) where the vertex of the mirror is located, then the coefficient D is 0, and distortion is fully corrected.
4. If spherical aberration and coma are both 0, the coefficient D is also 0.
5. If coma and astigmatism are both 0, distortion will be fully corrected.
6. For the elliptical mirror, the following aberrations cannot be corrected simultaneously:
   (a) astigmatism and field curvature;
   (b) spherical aberration, coma and astigmatism;
   (c) spherical aberration, coma and field curvature;
   (d) spherical aberration, astigmatism and distortion;
   (e) spherical aberration, field curvature and distortion.

6.3 Diffraction integral in the presence of aberration

Geometrical optics have a guiding significance on aberration correction, surface shape compensation and installation and commissioning of the elliptical mirror, but provides an approximate model, which is effectively only under the condition of short wavelength, so it can be expected that when the aberration decreases, geometrical optics will inevitably lose its effectiveness [1]. Meanwhile, when we discuss the focusing characteristics of the elliptical mirror with a high numerical aperture, it is necessary to analyze the diffraction theory of the aberration. Within the category of diffraction optics, the aberration makes the focusing wavefront after reflection by the elliptical mirror no longer an ideal spherical wave. Due to the machining error, the installation and commissioning error and the deviation of the point light source, aberrations will occur. Therefore, under the condition of high numerical aperture, the influence of the aberration becomes significant. For the elliptical mirror in the presence of aberration, the diffraction integral shall be corrected on the basis of the conclusion given in section 2.3.

6.3.1 Debye diffraction integral in the presence of aberration

According to literature [1], the focusing wavefront $\overline{W}$ after the reflection by the elliptical mirror is no longer an ideal spherical wave due to the aberration, but we still can make Gaussian reference sphere W through the center C of the diffraction aperture. Supposing the object point forms an image at the point P_0, a spherical surface is made with P_0 as the center and the distance between P_0 and C as the radius, thereby obtaining the Gaussian reference sphere W, the deviation of which from the actual focusing wavefront can represent the deformation degree of the wave surface in the diffraction aperture region, called wave aberration which is expressed by Φ. As shown in figure 6.2, P_1 and P_1^* are respectively intersections of some ray in the image space and the Gaussian reference sphere passing the point C and the actual wave surface, then

$$\Phi = P_1 P_1^*. \tag{6.22}$$

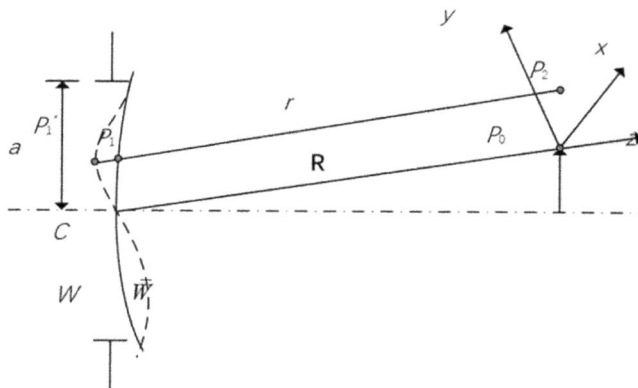

Figure 6.2. Focusing schematic diagram in the presence of aberration.

6-9

Supposing P_2 is near P_0, the light intensity distribution of the point P_1 on the Gaussian reference sphere can be expressed as follows, according to the derivation process of Debye integral in section 2.4

$$U(P_1) = \frac{P(P_1)\exp(ikR)}{R}. \tag{6.23}$$

Formula (6.24) represents a focusing spherical wave with P_0 as the center, wherein $P(P_1)$ has the same definition as the pupil function $P(r)$ in chapter 3. Under normal conditions, the wave aberration is the order of wavelength, so the light field distribution at the point P_1 can be used to approximately describe that at the point P_1^*.

Due to the presence of the wave aberration Φ, the light field distribution of spherical wavelets emitted from the point P_1^* at the observation point P_2 can be corrected as

$$\frac{\exp[-ik(r+\Phi)]}{r+\Phi}. \tag{6.24}$$

Here r represents the distance between $P_1(P_1^*)$ and P_2, and $k\Phi$ represents phase change caused by the wave aberration. Obviously, r is much more than the wave aberration Φ of the wavelength scale, so in practical application, Φ is negligible in the denominator of formula (6.25) which is substituted in formula (2.38), thereby obtaining the light field distribution of the region near the point P_2 in the presence of aberration, which is expressed by the integral on the Gaussian reference sphere.

$$U(P_2) = \frac{i}{\lambda} \iint_{\Sigma} P(P_1) \exp(-ik\Phi) \frac{\exp[ik(R-r)]}{Rr} \cos(\mathbf{n} \cdot \mathbf{r}) \, dS. \tag{6.25}$$

Based on the conclusion in section 2.3, formula (6.26) can be simplified with Debye approximation, thereby obtaining

$$U(r_2, \psi, z_2) = \frac{i}{\lambda} \iint_{\Omega} P(\theta) \exp(-ik\Phi)$$
$$\exp[-ikr_2 \sin\theta \cos(\varphi - \psi) - ikz_2 \cos\theta]\sin\theta d\theta d\varphi \tag{6.26}$$

where, $P(\theta)$ is the apodization factor of the elliptical mirror. Supposing γ is the maximum focusing angle, and the normalized radius $\rho = \theta/\gamma$, the light field distribution near the focal point of the elliptical mirror with a high numerical aperture in the presence of aberration is corrected as

$$U(v, \psi, u) = \frac{i\gamma}{\lambda} \int_0^1 \int_0^{2\pi} P(\rho) \exp(-ik\Phi)$$
$$\exp\left[-iv\frac{\sin(\rho\gamma)}{\sin\gamma}\cos(\varphi-\psi) - iu\frac{\cos(\rho\gamma)}{\sin^2(\gamma/2)}\right] \sin(\rho\gamma) d\rho d\varphi. \tag{6.27}$$

6.3.2 Strehl intensity

It should be noted that in the presence of aberration, the zero point ($u = 0$, $v = 0$) of the test surface may not be the point with the maximum light intensity. To describe this phenomenon, the concept of Strehl intensity is introduced by reference to the relevant monograph [1].

When the aberration does not exist, the light field intensity of the zero point is at its maximum. Supposing rays enter from an ideal point light source, and $P(\rho)$ represents the apodization factor of the elliptical mirror expressed by the normalized radius, then the maximum light intensity is expressed as

$$I_{wa} \propto |U(v = 0, u = 0)|^2 = k^2\gamma^2 \int_0^1 P(\rho)P^*(\rho)\sin^2(\rho\gamma)\, d\rho. \tag{6.28}$$

The light intensity distribution I is

$$I(v, \psi, u) = |U(v, \psi, u)|^2.$$

The normalized intensity is expressed as

$$i(v, \psi, u) = \frac{I}{I_{wa}} \tag{6.29}$$

where, the maximum $i(v,\psi,u)$ is called Strehl intensity. The space coordinate corresponding to the maximum $i(v,\psi,u)$ is called diffraction focus. Generally speaking, there may be more than one diffraction focus, but if the aberration is the order of wavelength, there is only one diffraction focus.

6.4 Zernike circle polynomial expansion of aberration function

When the aberration was discussed in geometrical optics, the aberration function Φ was expanded into a power series form, while in diffraction optics, as shown in formula (6.28), it is necessary to integrate a unit circle, so it should be considered that Φ is expanded into an orthogonal multinomial complete set in a unit circle. The multinomial complete set with this property may have various constitution forms; however, the Zernike circle polynomial has both orthogonality and simple invariance. More importantly, there is one-to-one correspondence between the Zernike circle polynomial and Seidel aberration [1]. Before the discussion of the Zernike circle polynomial expansion of the aberration function, the transference theorem is discussed firstly.

6.4.1 Transference theorem

Supposing Φ and Φ' are aberration functions, the relationship between them is as follows:

$$\Phi' = \Phi + H\rho^2 + K\rho \sin\theta + L\rho \cos\theta + M \tag{6.30}$$

where, H, K, L and M are constants in an order of magnitude of λ, and the transference theorem means the relationship met by Strehl intensity when there are aberration functions Φ and Φ' in the optical system.

$$i(v, \psi, u) = i(v', \psi', u').$$ (6.31)

That is to say, when an aberration function is added with $H\rho^2 + K\rho \sin \theta + L\rho \cos \theta + M$, the influence of this aberration function on the three-dimensional light field distribution near the focal point does not change, but just focusing spots are displaced in whole [1]. The specific displacement relationship is

$$\begin{cases} u' = u + 2kH \\ v' \cos \psi' = v \cos \psi - kK \\ v' \sin \psi' = v \sin \psi - kL \end{cases}$$ (6.32)

which respectively correspond to axial displacement and transverse displacement of diffraction light spot.

6.4.2 Zernike circle polynomial

The derivation of the Zernike circle polynomial is not repeated here, and its concrete form in a unit circle is given in the following formula:

$$R_n^{\pm m} (\rho) = \sum_{s=0}^{\frac{n-m}{2}} (-1)^s \frac{(n - s)!}{s! \left(\frac{n+m}{2} - s\right)! \left(\frac{n-m}{2} - s\right)!} \rho^{n-2s}$$ (6.33)

where, m, n and l are non-negative integers, and $m = |l|$. The aberration function Φ can be expanded into the form of Zernike circle polynomial. According to the symmetry, an image is formed at an ideal point. If the distance between the image point and the optical axis is Y_0, the variables in the expansion only have the combining forms of Y_0, ρ^2 and $Y_0\rho \cos \theta$, so the expansion will inevitably be presented in the following form:

$$\Phi(Y_0, \rho, \varphi) = \sum_l \sum_n \sum_m a_{lmn} Y_0^{2l+m} R_n^m (\rho) \cos (m\varphi)$$ (6.34)

where $n \geqslant m$, and n–m is an even number, and each a is an integer.

Here, the major concern is the diffraction image of a fixed object point, so the distance Y_0 from the location of the image point to the optical axis is a constant. Therefore, formula (6.35) can be rewritten as:

$$\Phi(\rho, \varphi) = A_{00} + \frac{1}{\sqrt{2}} \sum_{n=2}^{\infty} A_{n0} R_n^0 (\rho) + \sum_{n=1}^{\infty} \sum_{m=1}^{n} A_{nm} R_n^m (\rho) \cos (m\varphi).$$ (6.35)

When $m = 0$, the aberration function has nothing to do with the variable φ, so formula (6.36) can be written as

$$\Phi(\rho) = \sum_n A_{n0} R_n^0(\rho). \tag{6.36}$$

This expression is called spherical aberration function, wherein n represents the order of the spherical aberration.

The most prominent advantage of expanding the aberration function with Zernike circle polynomial is aberration 'balance' which means that aberrations of different levels are balanced each other to maximize the intensity of Gaussian focus. This conclusion is discussed in details in literature.

6.5 Primary aberration and its influence on the focusing characteristic of the elliptical mirror

This section expresses each primary aberration (Seidel aberration) as Zernike circle polynomial, and respectively studies the influence of each aberration on the focusing characteristic of the elliptical mirror.

The primary aberration corresponds to the terms in formula (6.36) meeting $2l + m + n = 4$, so its algebraic expression is as follows:

$$\Phi(\rho, \varphi) = \varepsilon_{nm} A_{lmn} R_n^m (\rho) \cos (m\varphi) \tag{6.37}$$

or

$$\Phi(\rho, \varphi) = A_{lmn} \rho^n \cos^m \varphi. \tag{6.38}$$

Here the value of ε_{mn} is

$$\varepsilon_{mn} = \begin{cases} 1 & m \neq 0 \\ 1/\sqrt{2} & m = 0, n \neq 0 \end{cases}. \tag{6.39}$$

The primary aberrations obtained in section 4.1 will be respectively discussed as follows.

6.5.1 Primary spherical aberration

Primary spherical aberration is also called fourth-order spherical aberration, and its parameter values are respectively $l = 0$, $n = 4$ and $m = 0$ [1]. It can be seen from the coefficient $l = m = 0$ that the primary spherical aberration is axially symmetrical, and its aberration function expression is

$$\Phi(\rho) = \frac{1}{\sqrt{2}} A_{040} R_4^0 (\rho) = \frac{1}{\sqrt{2}} A_{040}(6\rho^4 - 6\rho^2 + 1). \tag{6.40}$$

Figure 6.3 shows the light field distribution of the primary spherical aberration. It should be noted that formula (6.40) also contains a defocus amount and a constant term, but according to the transference theorem, they have no influence on the light

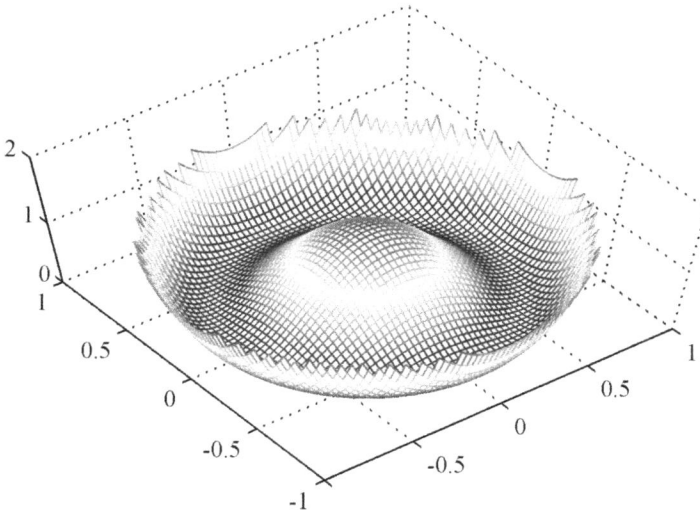

Figure 6.3. Schematic diagram of light field distribution of primary spherical aberration.

intensity distribution near the focal point, so formula (6.40) can be written into the following form:

$$\Phi(\rho) = A_{040}\rho^4. \tag{6.41}$$

The aberration function of the primary spherical aberration is substituted into formula (6.28), obtaining the light field distribution as shown in figure 6.4. The primary spherical aberration is rotationally symmetrical about the optical axis, so only the light field distribution of the v–u plane is discussed. It can be seen from the figure that when the aberration coefficient is small, and $A_{040} = 0.25\lambda$, as shown in figures 6.4(a) and (b), diffraction light spots translate along the positive direction of the optical axis. Then the diffraction light spots are different from those in the condition when there is no spherical aberration. However, when the aberration coefficient A_{040} further increases, as shown in figures 6.4(c) and (d), $A_{040} = 0.5\lambda$, the diffraction focus translates obviously, and the spherical aberration has significant influence on the focusing characteristic of the elliptical mirror.

Figure 6.5 quantitatively shows the change of the focusing characteristic curve of the elliptical mirror in the presence of spherical aberrations of different magnitudes. It can be seen that when the primary spherical aberration is small, the lateral response curve decreases only in its amplitude, but broadens obviously in its main lobe with the increase of the spherical aberration coefficient. But for the axial response, when the aberration coefficient is 0–0.5λ, the amplitude of the focusing light spots decreases slightly, and the diffraction focus translates axially, but the shape of the main lobe remains basically unchanged. However, when the aberration coefficient is 1λ, the diffraction light spots translate obviously, the main lobe broadens significantly, and then the contrast of detection signals decreases greatly.

a) Three-Dimensional Diagram of Light Field Distribution on Meridian Plane with $A_{040}=0.25\lambda$

b) Contour Diagram of Light Field of Meridian Plane with $A_{040}=0.25\lambda$

c) Three-Dimensional Diagram of Light Field Distribution on Meridian Plane with $A_{040}=0.50\lambda$

d) Contour Diagram of Light Field of Meridian Plane with $A_{040}=0.50\lambda$

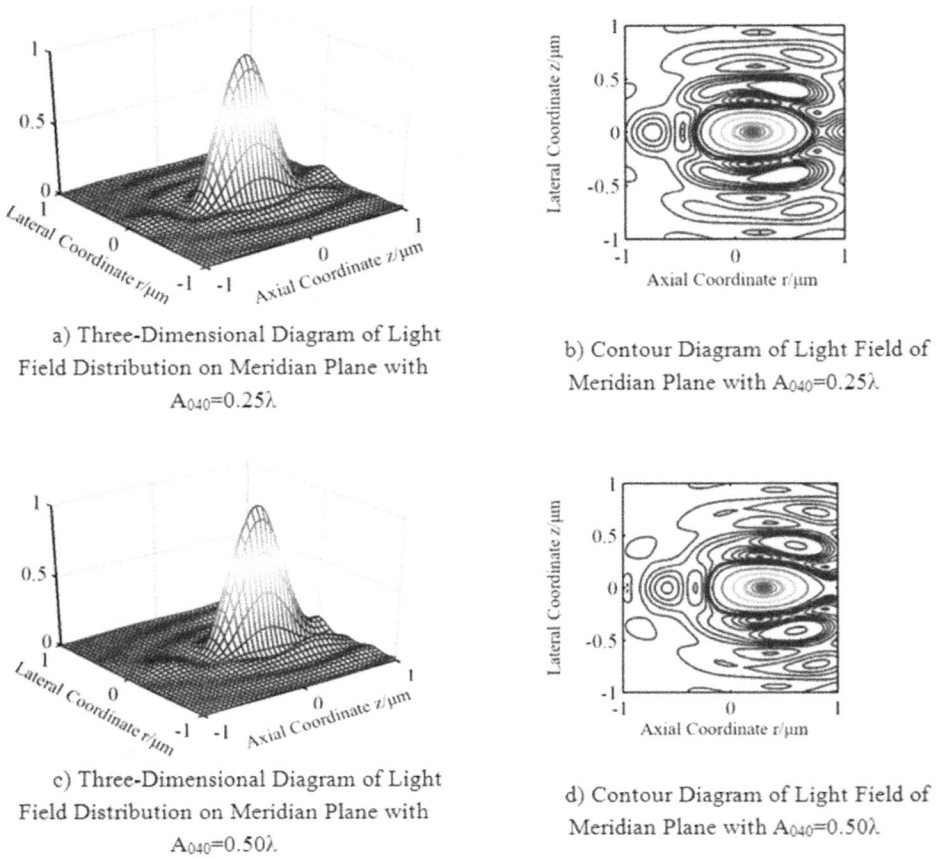

Figure 6.4. Light field distribution near focal point of elliptical mirror in the presence of primary spherical aberration.

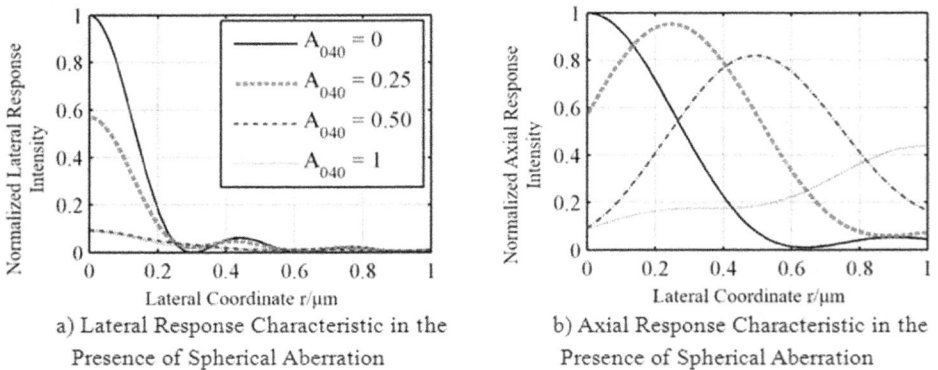

a) Lateral Response Characteristic in the Presence of Spherical Aberration

b) Axial Response Characteristic in the Presence of Spherical Aberration

Figure 6.5. Influence of primary spherical aberrations of different magnitudes on the focusing light spot of the elliptical mirror.

6.5.2 Primary coma

The parameter values of the primary coma are $l = 0$, $n = 3$ and $m = 1$ [1], which are substituted into formula (6.53), obtaining the aberration function expression of the primary coma as follows:

$$\Phi(\rho, \varphi) = A_{031}\rho^3 \cos \varphi. \tag{6.42}$$

The three-dimensional diagram of its light field distribution is shown in figure 6.6.

Although the expression of the primary coma contains $\cos \varphi$, the primary coma is still symmetrical about x axis (or y axis). The primary coma is no longer laterally and rotationally symmetrical, so the light intensity distributions of its focal plane and meridian plane are respectively discussed as follows.

It can be seen from figure 6.7 that on the focal plane of the elliptical mirror, when the primary coma coefficient is 0.25λ, the diffraction light spots are distorted slightly compared with the condition when there is no aberration, but the diffraction focus translates slightly along the negative direction of x axis, and for the reflective system, this direction is opposite to the translation direction of the diffraction focus when there is a primary coma in the lens system; when the primary coma coefficient is 0.5λ, the diffraction focus translates more obviously, and it is clear that the intensity of the slide lobe in the negative direction of x axis is apparently higher than that on the other side of the diffraction light spot.

Figure 6.8 shows the light field distribution of the meridian plane of elliptical mirror in the presence of primary coma of the corresponding magnitude. At this moment, the light spot is no longer prolate and out of shape. The larger the aberration coefficient, the more obvious the deformation. But obviously, the diffraction focus is always on the optical axis.

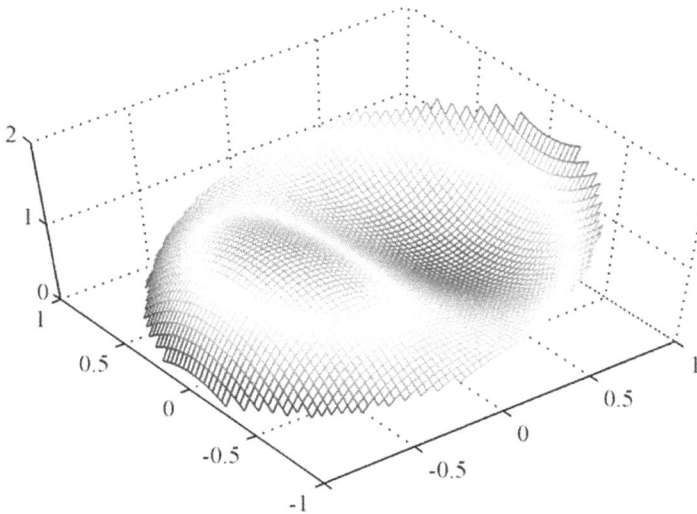

Figure 6.6. Schematic diagram of light field distribution of primary coma.

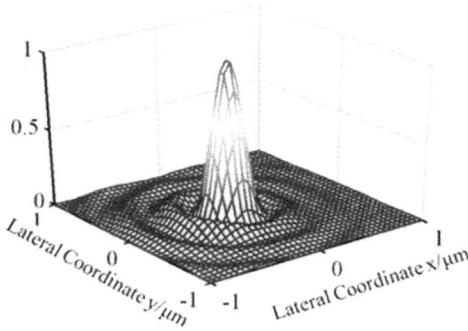

a) Three-Dimensional Diagram of Light Field Distribution on Focal Plane with $A_{031}=0.25\lambda$

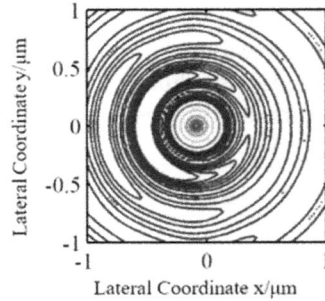

b) Contour Diagram of Light Field of Focal Plane with $A_{031}=0.25\lambda$

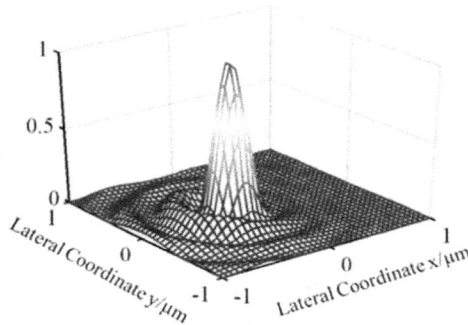

c) Three-Dimensional Diagram of Light Field Distribution on Focal Plane with $A_{031}=0.50\lambda$

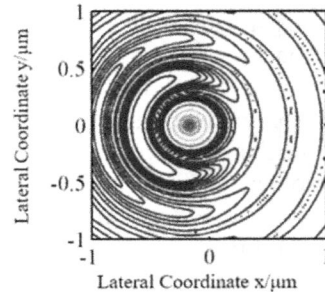

d) Contour Diagram of Light Field of Focal Plane with $A_{031}=0.50\lambda$

Figure 6.7. Light field distribution on focal plane of elliptical mirror in the presence of primary coma.

It is similar to the analysis of primary spherical aberration that figure 6.9 quantitatively shows the curve of light field distribution on the focal plane in the presence of comas of different magnitudes.

It can be seen that with the increase of the aberration coefficient, the light intensity of the diffraction focus gradually decreases, but only when the aberration coefficient is 1λ, does the focusing light spot broaden significantly, which indicates that small primary coma has less influence on the focusing characteristic of the system.

In fact, a coma is an off-axis aberration, and has great influence on the optical beam scanning in the elliptical reflective confocal microscopic imaging system. During scanning on an objective table, the axial point is imaged, and the coma mainly results from location offset or diffraction limitation of point light source, surface shape machining error, etc.

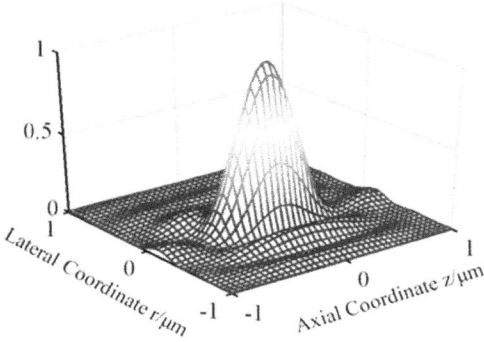

a) Three-Dimensional Diagram of Light Field Distribution on Meridian Plane with $A_{031}=0.25\lambda$

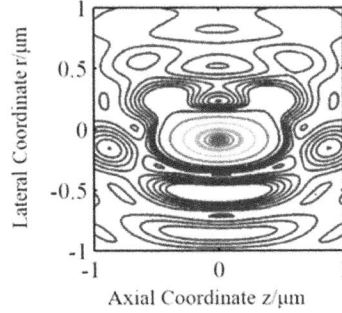

b) Contour Diagram of Light Field of Meridian Plane with $A_{031}=0.25\lambda$

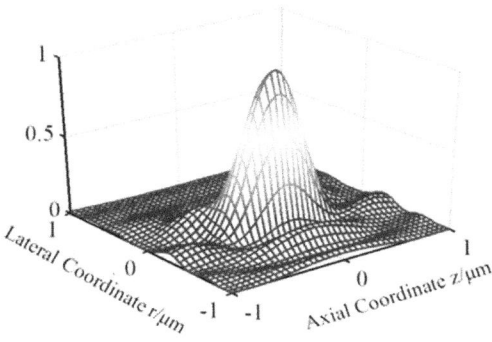

c) Three-Dimensional Diagram of Light Field Distribution on Meridian Plane with $A_{031}=0.50\lambda$

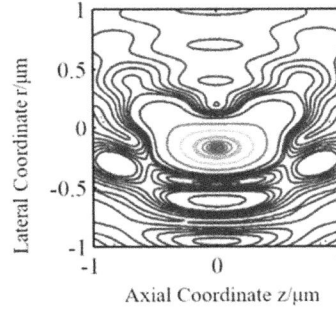

d) Contour Diagram of Light Field of Meridian Plane with $A_{031}=0.50\lambda$

Figure 6.8. Light field distribution on meridian plane of elliptical mirror in the presence of primary coma.

6.5.3 Primary astigmatism

The parameter values of the primary astigmatism are $l = 0$, $n = 2$ and $m = 2$ [1] and the aberration function expression of the primary astigmatism is:

$$\Phi(\rho,\ \varphi) = A_{022}\rho^2 \cos^2 \varphi. \tag{6.43}$$

The three-dimensional distribution of primary astigmatism is shown in figure 6.10. The primary astigmatism has complicated influence on light intensity distribution. In addition to the axial focal shift, diffraction light spot also broadens laterally on the focal plane, as shown in figures 6.11 and 6.12.

It can be seen from figure 6.11 that when the primary astigmatism coefficient A_{022} is 0.25λ, the diffraction light spots on the focal plane of the elliptical mirror are

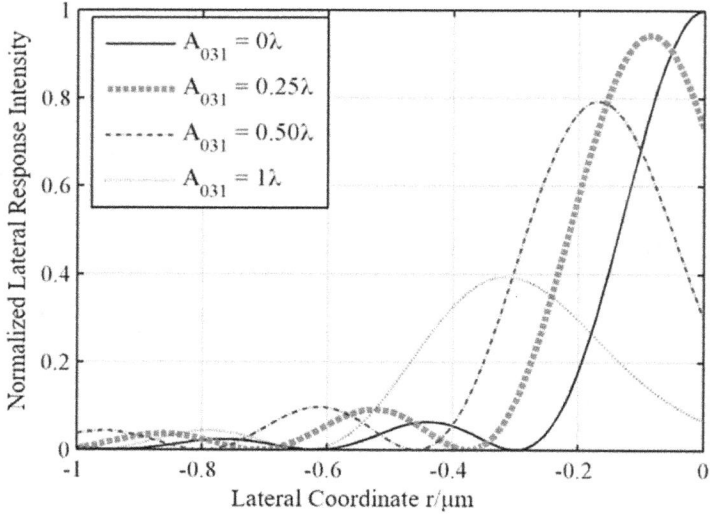

Figure 6.9. Influence of primary comas of different magnitudes on focusing light spot on focal plane of elliptical mirror.

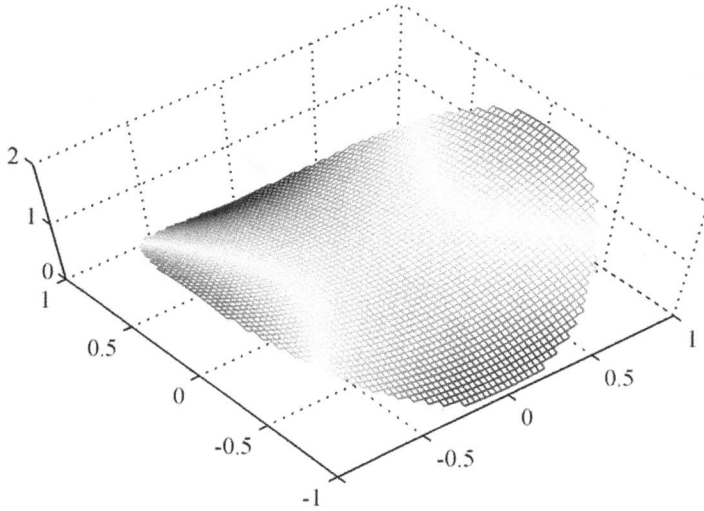

Figure 6.10. Schematic diagram of light field distribution of primary astigmatism.

basically unchanged, but when the astigmatism coefficient A_{022} is 0.5λ, the diffraction light spot broadens significantly in the lateral direction. At the same time, it can be seen that in the presence of the primary astigmatism, the light spot is only symmetrical respectively about x axis and y axis and no longer rotationally symmetrical, and then the amplitude of the side lobe increases significantly.

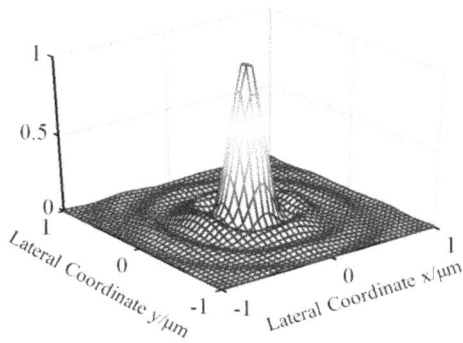

a) Three-Dimensional Diagram of Light Field Distribution on Focal Plane with $A_{022}=0.25\lambda$

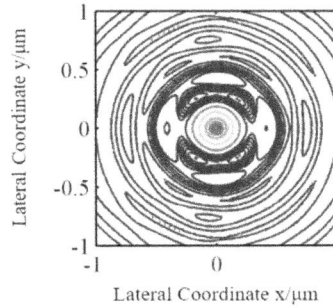

b) Contour Diagram of Light Field of Focal Plane with $A_{022}=0.25\lambda$

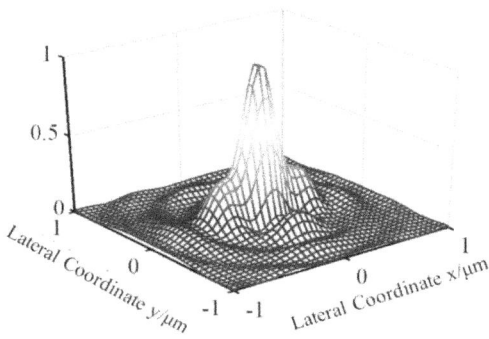

c) Three-Dimensional Diagram of Light Field Distribution on Focal Plane with $A_{022}=0.50\lambda$

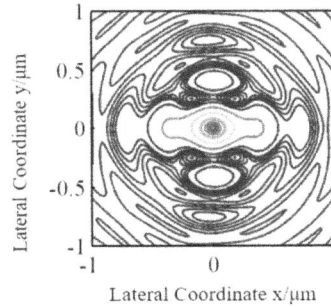

d) Contour Diagram of Light Field of Focal Plane with $A_{022}=0.50\lambda$

Figure 6.11. Light field distribution on the focal plane of the elliptical mirror in the presence of primary astigmatism.

However, the change of the light field distribution on the meridian plane in the presence of primary astigmatism is complex, as shown in figure 6.12. It can be seen from the change of lateral and axial light field distribution that in the presence of primary astigmatism, the diffraction focus is still on the optical axis ($x = 0$, $y = 0$) but translates along the axial direction of the optical axis.

We can see from the above three-dimensional and contour diagrams that the axial light field distribution is complex when the elliptical mirror has primary astigmatism, so we only discuss the lateral light field distribution of diffraction light spots in the presence of primary astigmatism, as shown in figure 6.13.

When $A_{022} = 0.5\lambda$, the main lobe and the side lobe of the response curve in the x direction cannot be significantly distinguished, and the image resolution of the optical system will be influenced at this moment, which indicates that the primary astigmatism has heavy influence on the focusing light spot of the elliptical mirror.

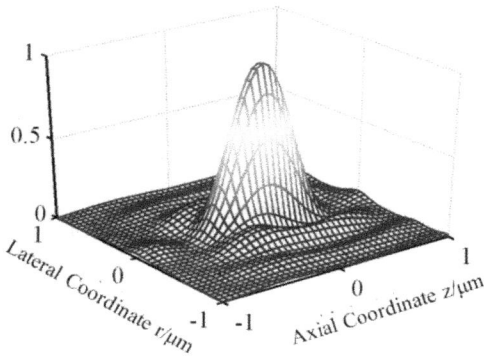

a) Three-Dimensional Diagram of Light Field Distribution on Meridian Plane with $A_{022}=0.25\lambda$

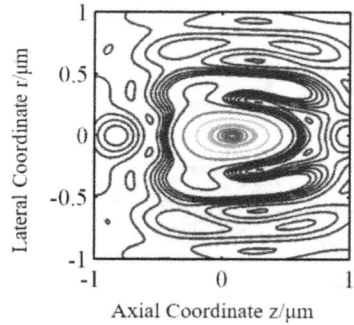

b) Contour Diagram of Light Field of Meridian Plane with $A_{022}=0.25\lambda$

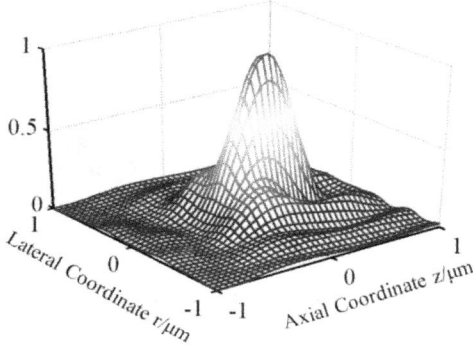

c) Three-Dimensional Diagram of Light Field Distribution on Meridian Plane with $A_{022}=0.50\lambda$

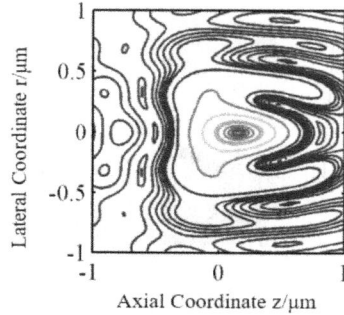

d) Contour Diagram of Light Field of Meridian Plane with $A_{022}=0.50\lambda$

Figure 6.12. Light field distribution on focal plane of elliptical mirror in the presence of primary astigmatism.

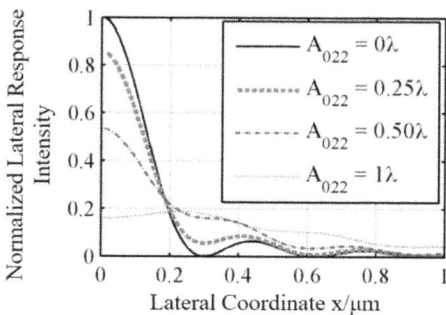

a) x-Direction Response Characteristics of Focal Plane

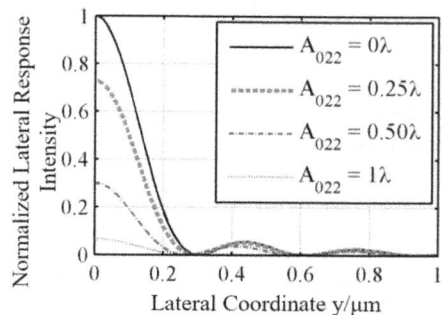

b) y-Direction Response Characteristics of Focal Plane

Figure 6.13. Influence of primary astigmatisms of different magnitudes on the focusing light spot of the elliptical mirror.

6.5.4 Field curvature and distortion

The parameter values of the field curvature are $l = 1$, $n = 2$ and $m = 0$ [1] and the aberration function expression of the field curvature is:

$$\Phi(\rho) = \frac{1}{\sqrt{2}} A_{120} R_2^0 (\rho) = \frac{1}{\sqrt{2}} A_{120}(2\rho^2 - 1). \qquad (6.44)$$

The parameter values of the distortion are $l = 1$, $n = 1$ and $m = 1$ [1] and the aberration function expression of the distortion is:

$$\Phi(\rho, \varphi) = A_{111} R_1^1 (\rho) \cos \varphi = A_{111}\rho \cos \varphi. \qquad (6.45)$$

The light field distributions of field curvature and distortion are shown in figures 6.14 and 6.15. According to the transference theorem, field curvature and distortion have no influence on the light intensity distribution of the diffraction light spot but have great influence on optical beam scanning of the confocal microscope.

Among the above five primary aberrations, according to the transference theorem, field curvature and distortion only cause the diffraction light spot to translate, and have no influence on the light intensity distribution near the focal point. Therefore, in topic analysis and the subsequent experiment and application, we need to pay close attention to spherical aberration, coma and astigmatism as well as their influences in processing of the elliptical mirror and design of the installation and commissioning system. Then, the maximum permissible aberration of the elliptical mirror system for each primary aberration is called aberration tolerance condition in Principles of Optics.

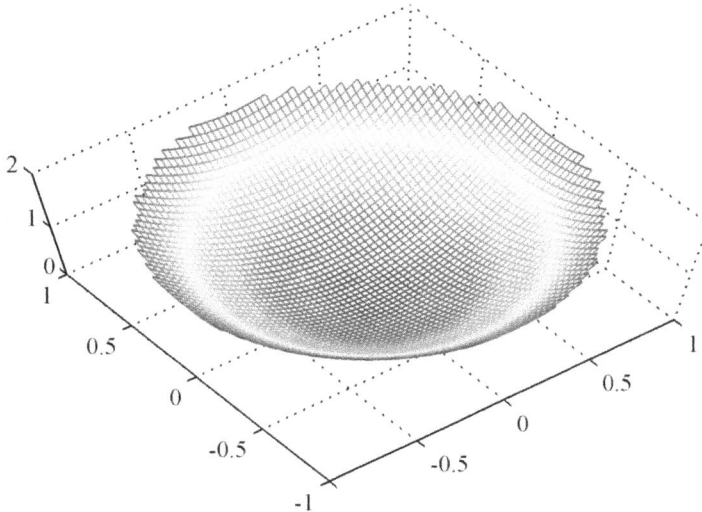

Figure 6.14. Schematic diagram of the light field distribution of field curvature.

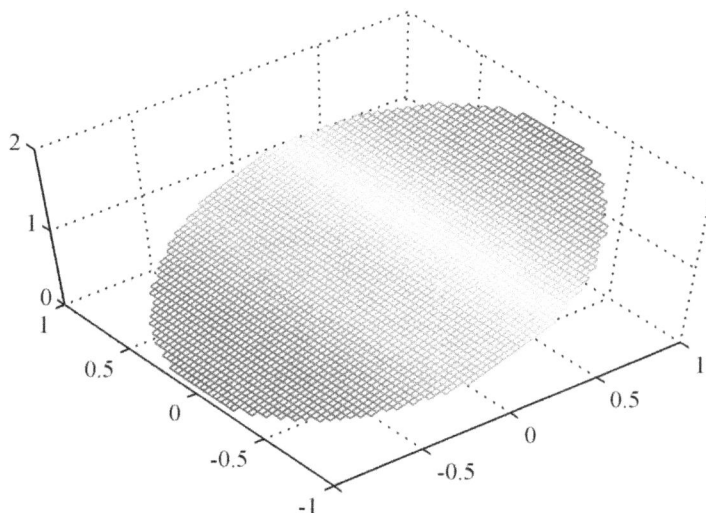

Figure 6.15. Schematic diagram of the light field distribution of distortion.

6.5.5 Aberration tolerance of elliptical mirror

It can be seen from the above analysis that the aberration changes the light intensity distribution near the focal point and therefore influences the image quality and even the image resolution. When evaluating the image quality of the optical imaging system, we need to know the tolerance condition of this system. We adopt Marechal's criterion. The relationship between the intensity of the reference sphere center and the root mean square of deviation of wave surface from the sphere is adopted in the criterion.

Marechal's criterion is elaborated as follows [1]: when the normalized intensity at the diffraction focus of an optical system is more than or equal to 0.8, the system will be regarded as a well-corrected system, that is:

$$i(v, \psi, u) = 0.8. \tag{6.46}$$

For different primary aberrations, the aberration tolerances are different. Figures 6.16–6.18 respectively show the relation curve of the normalized intensity and the aberration coefficient at the diffraction focus of the elliptical mirror system with primary spherical aberration, coma and astigmatism under different maximum focusing angles.

It can be obtained from the figures that when the numerical aperture is 1 (the maximum focusing angle $\gamma = \pi/2$), the aberration tolerance condition of the elliptical mirror is: for primary spherical aberration, $|A_{040}| \leqslant 0.52\lambda$; for primary coma, $|A_{031}| \leqslant 0.49\lambda$; for primary astigmatism, $|A_{022}| \leqslant 0.35\lambda$.

Primary field curvature and primary distortion are respectively expressed by quadratic term and linear term of ρ, so according to the transference theorem, their only influence is to result in the whole translation of three-dimensional aberration-less intensity distribution. Therefore, when there is less primary field curvature or

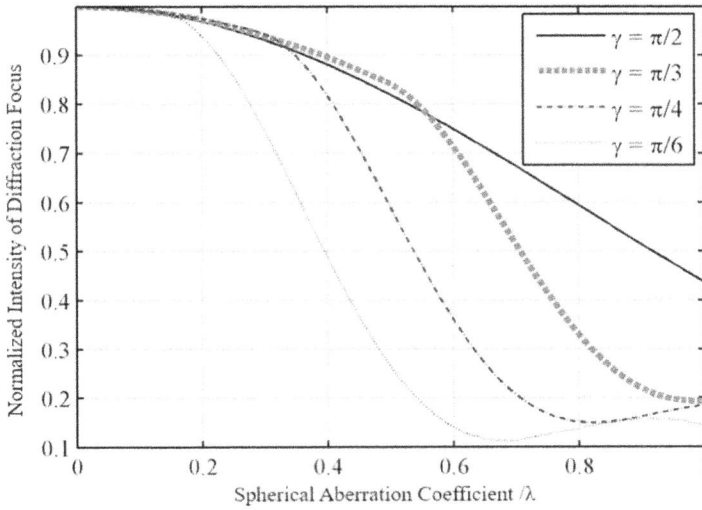

Figure 6.16. Normalized intensity curve at the diffraction focus of the elliptical mirror in the presence of primary spherical aberration.

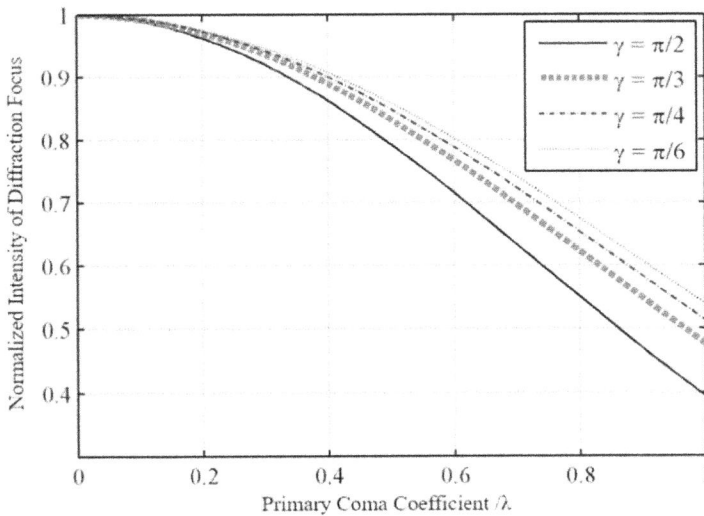

Figure 6.17. Normalized intensity curve at the diffraction focus of the elliptical mirror in the presence of primary coma.

primary distortion, the normalized intensity of the diffraction focus i is 1, but the diffraction focus does not coincide with the Gaussian image point.

According to the aberration tolerance conditions of the lens in literature [1], the aberration tolerance conditions of the elliptical mirror and the lens are listed in table 6.1, which indicates that the elliptical mirror and the lens are the most sensitive to

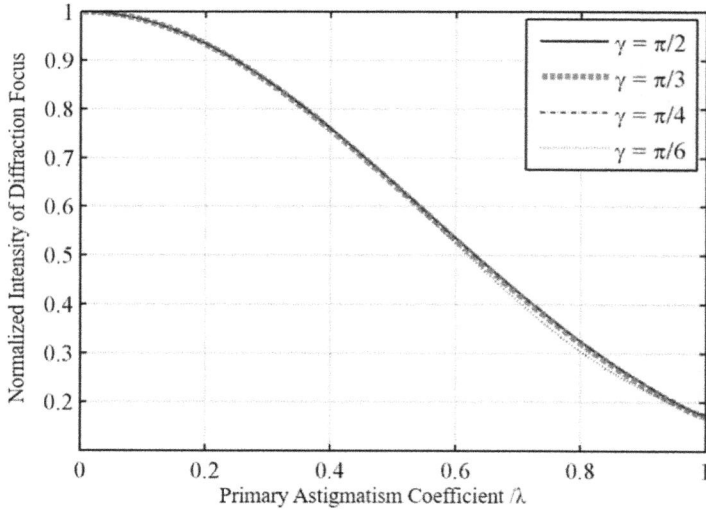

Figure 6.18. Normalized intensity curve at the diffraction focus of the elliptical mirror in the presence of primary astigmatism.

Table 6.1. Aberration tolerance conditions of elliptical mirror and lens.

	Elliptical mirror	Lens [1]
Spherical Aberration	0.52λ	0.94λ
Coma	0.49λ	0.60λ
Astigmatism	0.35λ	0.35λ

primary astigmatism. Accordingly, the elliptical mirror is more sensitive to primary spherical aberration and coma than the lens, which is one of the disadvantages of the reflective system compared with the lens system.

6.6 Conclusion

This chapter analyzes the aberration of the mirror with a high numerical aperture based on geometrical optics, and gives the expression of each primary aberration coefficient under certain surface shape parameter conditions. This chapter also studies the aberration characteristics of the elliptical mirror based on the scalar diffraction theory, establishes a diffraction integral of the elliptical mirror with a high numerical aperture in the presence of the aberration, expands the aberration function with the Zernike circle polynomial and particularly analyzes the influence of primary spherical aberration, coma and astigmatism on the focusing character-istic to provide a necessary theoretical basis for the processing of the elliptical mirror and the design of the installation and commissioning system. Through simulation analysis, the tolerance conditions of the elliptical mirror for each primary aberration are obtained: primary spherical aberration $|A_{040}| \leqslant 0.52\lambda$, primary coma $|A_{031}| \leqslant 0.49\lambda$ and primary astigmatism $|A_{022}| \leqslant 0.35\lambda$, which are also the important basis for determining the machining precision parameter of the elliptical mirror which is a core optical element.

References

[1] Born M and Wolf E 2013 *Principles of Optics: Electromagnetic Theory of Propagation, Interference and Diffraction of Light* (Amsterdam: Elsevier)

[2] Korsch D 2012 *Reflective Optics* (New York: Academic)

[3] Liu J *et al* 2013 Focusing of an elliptical mirror based system with aberrations *J. Opt.* **15** 105709

Chapter 7

Three-dimensional transfer function

Shan Gao, Mengzhou Li, Tong Wang, Jian Liu and Jiubin Tan

7.1 Introduction

Transfer function is widely used to investigate the imaging performance of an optical system. As we know, the objective can be resolved to series of spatial frequency components, and the efficiency with which various spatial frequencies are transmitted is the transfer function. The larger scale of the transfer spatial frequencies transmits the system, and the performance of the image will be better [1].

In this chapter, we analyze the transfer function of the elliptical mirror because it is a mathematical expression for imaging description of the elliptical system and also the basic theory for the elliptical mirror of confocal scanning microscopic imaging. We will study the point spread function, the coherent transfer function and the optical transfer function of an elliptical mirror, and will use the three-dimensional transfer function theory to study the elliptical reflective confocal microscopic system.

This chapter begins with the point spread function to analyze the coherent transfer function and the optical transfer function of the elliptical mirror in section 7.2. In section 7.3, we will establish the three-dimensional transfer function model of the elliptical mirror with a high numerical aperture.

7.2 Point spread function

When the elliptical mirror is used in the imaging system or confocal scanning system, it is necessary to analyze its imaging characteristics, and the most commonly used characterization methods are point spread function and optical transfer function.

For the imaging of the lens system, we know that the unit impulse function of the point light source P_1 is a δ function. However, the image formed by P_1 passing the optical system on the image plane is no longer a δ function due to diffraction and aberration caused by the finite aperture of the optical system, but a certain scale of distribution of light intensity or complex amplitude, expressed as $h(u, v)$ which is called an impulse response function. In particular, for the imaging of an incoherent point light source, $h(u, v)$ is called the point spread function. This section is mainly

doi:10.1088/978-0-7503-1629-3ch7

based on the microscopic imaging system for coherent illumination, but the impulse response function and the point spread function for incoherent illumination are still collectively called point spread function.

The elliptical mirror is discussed as follows by reference to the analysis method of M Gu *et al* for the point spread function and transfer function of the lens with a high numerical aperture [2]. According to the definition of the point spread function, we can obtain the point spread function of the elliptical mirror axis at the perifocus P_2 based on the analysis of focusing characteristics in section 5.4:

$$h_{P_1-P_2}(v_2, u_2) = \exp\left[-\frac{iu_2}{4\sin^2(\gamma/2)}\right]$$
$$\times \int_0^\gamma P(\theta)\, J_0\left(\frac{v_2\sin\theta}{\sin\gamma}\right)\exp\left(\frac{iu_2\sin^2(\theta/2)}{2\sin^2(\gamma/2)}\right)\sin\theta d\theta. \tag{7.1}$$

Lateral and axial optical coordinates v and u near the point P_2 are respectively:

$$v_2 = \frac{2\pi}{\lambda}r_2\sin\gamma$$

$$u_2 = \frac{8\pi}{\lambda}z_2\sin^2\left(\frac{\gamma}{2}\right)$$

where $P(\theta)$ is an apodization factor for the rotating elliptical mirror in the case of perfect point illumination $(P(\alpha) = 1)$;

$$P(\theta) = \left|\frac{a+c}{a-c}\times\frac{\sin[g(\theta)]}{\sin\theta}g'(\theta)\right|. \tag{7.2}$$

$g(\theta)$ is a mapping function, with the form as shown in formula (5.9).

Accordingly, for the elliptical mirror shown in figure 5.4, when P_2 is illuminated by a perfect point light source, and P_1 is the position where the image plane is, the angle θ is still used as an integral variable (θ is an object aperture angle), and the point spread function of the elliptical mirror is rewritten as

$$h_{P_2-P_1}(v_1, u_1) = \exp\left[-\frac{iu_1}{4\sin^2(\varphi/2)}\right]$$
$$\times \int_0^\gamma \frac{1}{P(\theta)} J_0\left(\frac{v_1\sin[g(\theta)]}{\sin\varphi}\right)\exp\left(\frac{iu_1\sin^2[g(\theta)/2]}{2\sin^2(\varphi/2)}\right)\sin[g(\theta)]g'(\theta)d\theta. \tag{7.3}$$

Here $g'(\theta)$ is the first derivative of the mapping function α-θ with respect to θ, and φ is the maximum focusing angle at P_1 and is concerned with the maximum value of the object aperture angle θ.

$$\varphi = \alpha_{max} = \arctan\left[\frac{(a^2-c^2)}{(a^2+c^2)\cos\gamma + 2ac}\right]. \tag{7.4}$$

It should be noted that formula (7.1) is different from the point spread function of the optical imaging system under paraxial approximation in that it does not have three-dimensional space invariance, so even a perfect rotating elliptical mirror can only be applied in objective table scanning.

7.2.1 Coherent transfer function

The coherent transfer function is three-dimensional Fourier transform of the point spread function:

$$c(\mathbf{m}) = \int_{-\infty}^{\infty} h(\mathbf{r})\exp(2\pi i \mathbf{r} \cdot \mathbf{m}) \, d\mathbf{r}. \tag{7.5}$$

Here $\mathbf{m}$ and $\mathbf{r}$ are respectively spatial frequency vector and position vector. For the elliptical mirror with a high numerical aperture, its three-dimensional point spread function under the Debye approximation condition is formula (5.16), which is substituted into formula (7.3) to obtain the three-dimensional coherent transfer function.

$$c(l, s) = K\int_0^{\infty} \int_0^{\infty} \exp\left[-\frac{iu}{4\sin^2(\gamma/2)}\right] J_0(2\pi l r_2)\exp(2\pi i z_2 s) r_2$$
$$\times \left[\int_0^{\gamma} P(\theta) J_0\left(\frac{v\sin\theta}{\sin\gamma}\right)\exp\left(\frac{iu\sin^2(\theta/2)}{2\sin^2(\gamma/2)}\right)\sin\theta \, d\theta\right] dr_2 dz_2 \tag{7.6}$$

where K is a normalized constant and can be ignored in calculation. Supposing $\sin\theta/\sin\gamma = \rho$, called the normalized coordinate of the elliptical mirror, formula (7.6) can be rewritten as

$$c(l, s) = K\int_0^{\infty} \int_0^{\infty} J_0(2\pi l r_2)\exp(2\pi i z_2 s) r_2$$
$$\times \left[\int_0^{\sin\gamma} \frac{P(\rho) J_0(kr_2\rho\sin\gamma)\exp\left(-ikz_2\sqrt{1 - \rho^2\sin^2\gamma}\right)}{\sqrt{1 - \rho^2\sin^2\gamma}} \rho d\rho\right] dr_2 dz_2. \tag{7.7}$$

Here $P(\rho)$ is the apodization factor with ρ as a variable, the mathematic relation

$$\int_0^{\infty} 2\pi J_0(2\pi r a) J_0(2\pi r b) \, r dr = \frac{\delta(a - b)}{a}$$

is used for executing integral operation on r_2, and then formula (7.7) can be expressed as

$$c(l, s) = K\int_0^{\infty} \left[\int_0^{\sin\gamma} \frac{P(\rho)\exp\left(-ikz_2\sqrt{1 - \rho^2\sin^2\gamma}\right)}{\sqrt{1 - \rho^2\sin^2\gamma}}\right.$$
$$\left. \times \delta\left(\frac{\rho\sin\gamma}{\lambda} - l\right) d\rho\right]\exp(2\pi i z_2 s) \, dz_2. \tag{7.8}$$

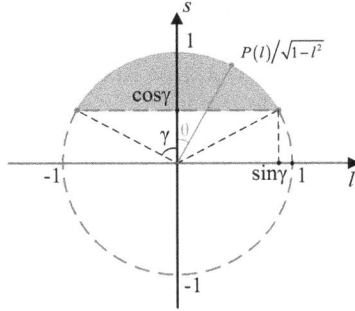

Figure 7.1. Three-dimensional coherent transfer function of the elliptical mirror with a high numerical aperture.

According to the properties of the δ function,

$$l = \frac{\rho \sin \gamma}{\lambda} = \frac{\sin \theta}{\lambda}. \tag{7.9}$$

λl and λs are respectively used for expressing l and s in the original formula without loss of physical meaning. Then $l = \sin \theta$, and finally, formula (7.8) can be expressed as

$$c(l, s) = \frac{P(l)}{\sqrt{1 - l^2}} \delta\left(s - \sqrt{1 - l^2}\right). \tag{7.10}$$

Here $P(l)$ is the apodization factor with l as a variable. It can be seen from formula (7.10) that the three-dimensional coherent transfer function of the elliptical mirror is a spherical cap, as shown in figure 7.1.

The spherical surface shown in figure 7.1 is usually known as the Ewald sphere, and the spherical equation is

$$s^2 + l^2 = 1.$$

The weight coefficient function of the Ewald spherical surface is $P(l)/\sqrt{1 - l^2}$. It can be seen from figure 7.1 that the lateral spatial cutoff frequency is $l = \sin \gamma$, the minimum axial spatial cutoff frequency is $s = \cos \gamma$, and axial translation also exists. Only when $\gamma = \pi/2$, is the minimum axial spatial cutoff frequency 0, and the axial translation disappears. Then the three-dimensional coherent transfer function of the elliptical mirror is the Ewald hemisphere.

It should be noted that when the maximum focusing angle γ is small, the optical system meets paraxial approximation, and then the Ewald sphere can be described with a paraboloid, which is commonly used in the lens system.

7.2.2 Optical transfer function of elliptical mirror

The optical transfer function and the coherent transfer function are respectively used for describing transfer functions of one imaging system in incoherent illumination and coherent illumination, which depend on the physical properties of the system. The optical transfer function of the elliptical mirror is discussed as follows.

The three-dimensional optical transfer function of the elliptical mirror with a high numerical aperture is the three-dimensional Fourier transform of the modular square of its point spread function. According to the convolution theorem for Fourier transform,

$$C(l, s) = \frac{P(l)}{\sqrt{1 - l^2}} \delta\left(s + \sqrt{1 - l^2}\right) \otimes_3 \frac{P(l)}{\sqrt{1 - l^2}} \delta\left(s - \sqrt{1 - l^2}\right). \qquad (7.11)$$

The optical transfer function of the elliptical mirror in incoherent illumination is the three-dimensional convolution of two spherical caps symmetric about the origin, as shown in figure 7.2.

For the three-dimensional optical function, there are many solution and expression methods. The most commonly used is the solution method directly using three-dimensional convolution, which has the disadvantages of complex programming and a large amount of calculation. Therefore, in this section, the three-dimensional convolution of two spherical shells is further simplified, thereby obtaining the explicit expression of the three-dimensional optical transfer function.

Supposing the normalized spatial frequencies in x, y and z directions are defined as m, n and s under the condition of one-wavelength laser source, and the lateral normalized spatial frequency is defined as l, then

$$l = \sqrt{m^2 + n^2}.$$

Accordingly, s is also called axial normalized spatial frequency.

The positions of sphere centers of two spherical surfaces are respectively defined as $(\pm|l|/2, \pm|s|/2)$ without loss of generality. Supposing P is a point on the

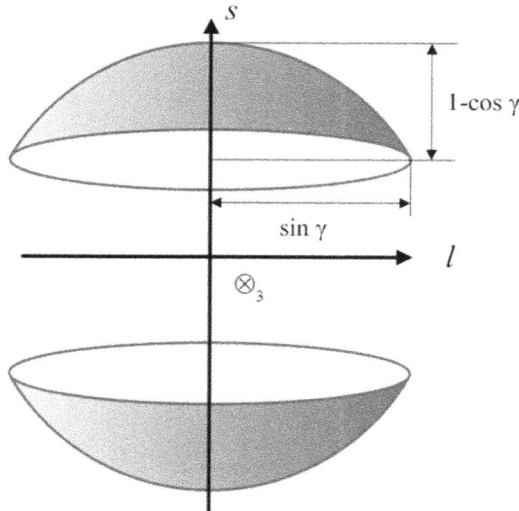

Figure 7.2. Convolution operation graph of three-dimensional optical transfer function.

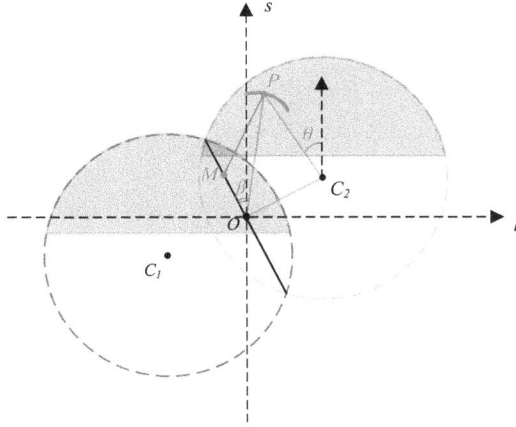

Figure 7.3. Convolution operation diagram of optical transfer function.

intersecting line of the two spherical shells, the local geometrical relationship of the point P on a plane through OM and perpendicular to MC_1C_2 is shown in figure 7.3, and its axial coordinate is

$$s' = -\frac{|l|}{\sqrt{l^2 + s^2}} \cos \beta \sqrt{1 - \frac{l^2 + s^2}{4}}. \tag{7.12}$$

Here β is the included angle between MO and PO, and then the optical transfer function of the elliptical mirror can be expressed as the integral form of the product of the apodization factor $P(\theta)$:

$$C(l, s) = \frac{4}{\pi\sqrt{l^2 + s^2}} \int_0^{\beta_1} P_1(\theta_1) P_2(\theta_2) \, d\beta \tag{7.13}$$

where the integral limit β_1 is the maximum angle of β on the intersecting line, and θ_1 and θ_2 are respectively the included angles between PC_1 and PC_2 and the negative axis s.

$$\cos \theta_1 = \frac{|l|}{\sqrt{l^2 + s^2}} \cos \beta \sqrt{1 - \frac{l^2 + s^2}{4}} - \frac{|s|}{2}$$

$$\cos \theta_2 = \frac{|l|}{\sqrt{l^2 + s^2}} \cos \beta \sqrt{1 - \frac{l^2 + s^2}{4}} + \frac{|s|}{2}. \tag{7.14}$$

To simplify the above formula, the following parameter is introduced:

$$p = \frac{2|l|}{|s|\sqrt{l^2 + s^2}} \sqrt{1 - \frac{l^2 + s^2}{4}}. \tag{7.15}$$

Therefore,

$$\cos \theta_{1,2} = \frac{|s|}{2}(p \cos \beta \mp 1). \qquad (7.16)$$

The other parameter β_1 in formula (7.13) depends on the maximum focusing angle, and can be expressed as

$$\beta_1 = \arccos\left[\frac{|s|/2 + \cos \gamma}{\frac{|l|}{\sqrt{l^2 + s^2}}\sqrt{1 - \frac{l^2 + s^2}{4}}}\right] = \arccos\left[\frac{1}{p}\left(\frac{2 \cos \gamma}{|s|} + 1\right)\right]. \qquad (7.17)$$

Formulae (7.13), (7.16) and (7.17) are used for numerical computation to obtain the three-dimensional optical transfer function of the elliptical mirror in incoherent illumination, as shown in figure 7.4.

It is found that there is a singular point at the position where $s = 0$ and $l = 0$, and under the condition of high numerical aperture, the optical transfer function of the elliptical mirror increases, and the corresponding imaging resolution also increases.

It can be quantitatively seen from figure 7.5 that when the maximum focusing angle of the elliptical mirror is $\pi/4$, the frequency component with $l > 1.5$ cannot be distinguished; when the maximum focusing angle is $\pi/3$, the numerical aperture has reached 0.85, and its lateral cutoff frequency is 1.75; when the maximum focusing angle is $\pi/2$, its lateral high-frequency component is obviously improved, and the cutoff frequency is 2. Accordingly, the cutoff frequency of the three-dimensional optical transfer function of the elliptical mirror is determined by the following formula:

$$2(|l|\sin \alpha - |s|\cos \alpha) \leqslant l^2 + s^2. \qquad (7.18)$$

Next, the three-dimensional optical transfer functions of the elliptical mirror and lens with a high numerical aperture are compared. Figure 7.6 shows the three-dimensional optical transfer function of the lens with a high numerical aperture.

Apparently, the lens is considered to be perfect. It can be found by comparing figure 7.4 with figure 7.6 that under the condition of the same numerical aperture,

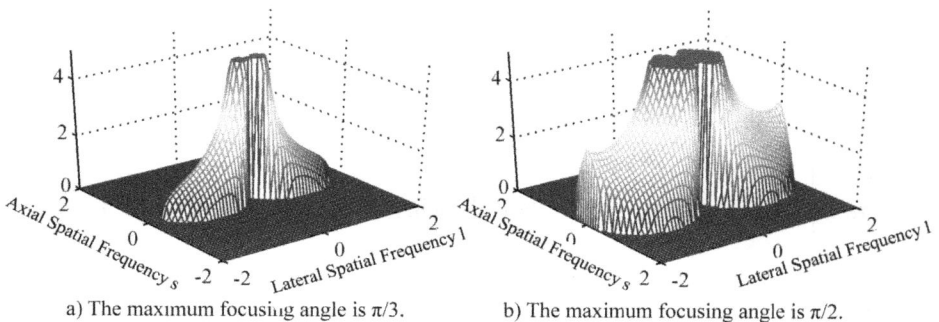

a) The maximum focusing angle is $\pi/3$.   b) The maximum focusing angle is $\pi/2$.

Figure 7.4. Three-dimensional optical transfer function of the elliptical mirror.

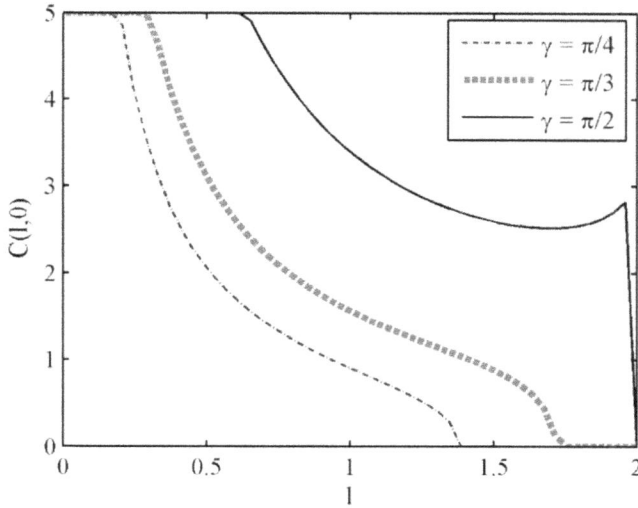

Figure 7.5. Three-dimensional optical transfer function of elliptical mirror at different maximum focusing angles.

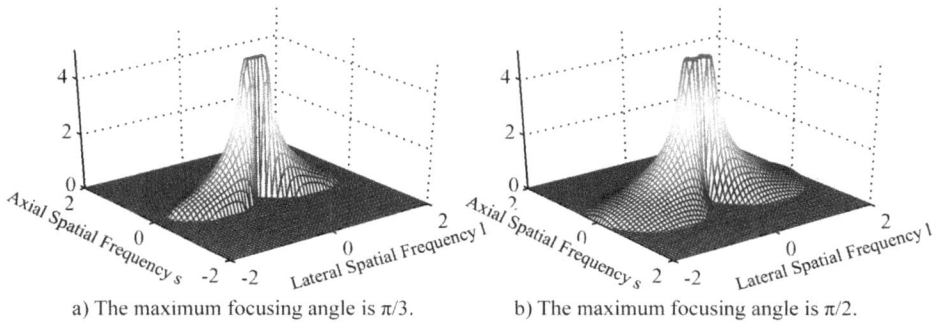

a) The maximum focusing angle is $\pi/3$. b) The maximum focusing angle is $\pi/2$.

Figure 7.6. Three-dimensional optical transfer function of lens with a high numerical aperture under the Sine condition.

when the numerical aperture is small, the optical transfer functions of the elliptical mirror and the lens have little difference; when the numerical aperture is 1.0, the optical transfer function of the lens slightly increases in response at high frequency, and the optical transfer function of the elliptical mirror significantly increases. This also reflects that the elliptical mirror is better than the lens in resolving power, particularly in high-frequency information distribution.

7.3 Three-dimensional transfer function of an elliptical reflective confocal microscopic system

Section 7.1 describes the imaging characteristics of the elliptical mirror from the perspective of the point spread function and the transfer function. This section will focus on the three-dimensional transfer function of the elliptical reflective confocal

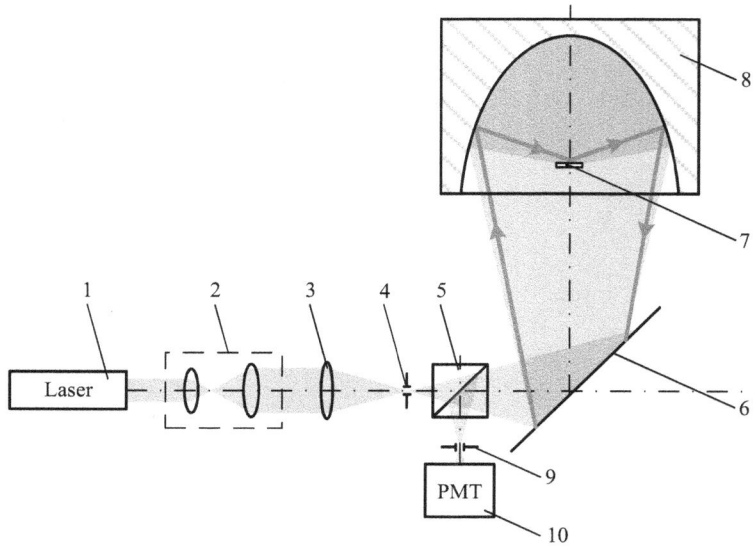

1. Laser 2. Collimating Beam Expander 3. Focusing Objective with High Numerical Aperture
4. Pinhole A 5. Beam Splitter 6. Plane Mirror 7. Sample 8. Elliptical Mirror 9. Pinhole B 10.
Photodetector

Figure 7.7. Schematic diagram of an elliptical reflective confocal imaging system.

microscopic imaging system based on the existing theory. It should be noted that only the coherent transfer function of the elliptical reflective confocal microscope under the conditions of point illumination and point detection is discussed in this section, and the field range of the system is restricted to the lighting point. Figure 7.7 shows the optical system diagram of the elliptical reflective confocal microscopic imaging system.

The elliptical reflective confocal microscopic imaging system adopts coherent illumination and has the similar structure to the confocal microscopic system consisting of a traditional lens. The traditional lens is replaced with the elliptical mirror, which can realize illumination and detection with the numerical aperture of 1 in order to make a breakthrough in lateral and axial resolution, but the imaging contrast of the low-frequency component also declines. This will be analyzed below from the perspective of the theory of the three-dimensional coherent transfer function.

7.3.1 Coherent transfer function of elliptical reflective confocal microscopic system

As shown in figure 7.7, under the ideal condition, the confocal microscopic system is in point illumination and point detection. Therefore, the three-dimensional point spread function $h_a(v, u)$ of the system can be expressed as the product of the lens point spread function h_1 and the detection point spread function h_2, wherein the point spread function of the lens in the traditional confocal system is replaced by the point spread function of the elliptical mirror.

$$h_a(v, u) = h_1(v, u)h_2(v, u). \tag{7.19}$$

It should be noted that in the optical imaging system, the analysis of space invariance is an important link of the theory system. Particularly as the confocal microtechnique gradually requires imaging in a larger field of view to enhance the scanning rate, space invariance gets more attention in instrument design. For the space invariance theory, Prof. M Gu and his team have made outstanding contribution [1]. It can be considered that space invariance is a space response characteristic of the optical system under limited conditions, and aberration needs perfect correction under the condition of a large field of view, which is extremely strict. Compared with a traditional transmissive system, the reflecting system is more sensitive to aberration. Especially, it can be seen from the analysis conclusion in chapter 6 that under the condition of high numerical aperture, the aberration function has the characteristics of more complex form and more obvious influence, so the elliptical mirror system has space invariance within a small field of view only in theory.

However, if objective or objective table scanning is adopted in the system, confocal imaging still can be expressed as a convolution of a three-dimensional object function and a three-dimensional point spread function $h_a(v, u)$ of system, because the optical properties of the confocal system remain unchanged at this moment, and then the elliptical reflective confocal microscopic system has a three-dimensional space invariance. Therefore, the optical beam scanning mechanism is not included in this section. During objective or objective table scanning, the light field distribution of the detector is

$$I(v_x, v_y, u) = |h_a(v, u) \otimes_3 o(v_x, v_y, u)|^2. \tag{7.20}$$

Here, $o(v_x, v_y, u)$ is the three-dimensional object function, with the lateral optical coordinates v_x and v_y and the axial optical coordinate u as variables.

Accordingly, the three-dimensional coherent transfer function of the elliptical reflective confocal microscopic imaging system can be expressed as the three-dimensional inverse Fourier transform of formula (7.19)

$$c(l, s) = \mathbb{F}_3[h_1(v, u)] \otimes_3 \mathbb{F}_3[h_2(v, u)]. \tag{7.21}$$

The confocal system shown in figure 7.7 is generally a reflective confocal system, and the elliptical mirror is used as an objective and has the functions of collection and detection. Therefore

$$h_1(v, u) = h(v, u)$$
$$h_2(v, u) = h(v, u).$$

$h(v,u)$ is substituted into formula (7.21), thereby getting

$$c(l, s) = \frac{P(l)}{\sqrt{1 - l^2}} \delta\left(s + \sqrt{1 - l^2}\right) \otimes_3 \frac{P(l)}{\sqrt{1 - l^2}} \delta\left(s + \sqrt{1 - l^2}\right). \tag{7.22}$$

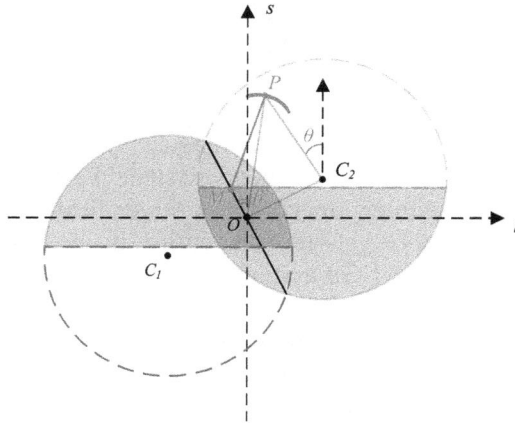

Figure 7.8. Schematic diagram of convolution operation for solving the coherent transfer function of a confocal system.

Formula (7.22) corresponds to the self-convolution of a spherical shell. Obviously, the coherent transfer function of the elliptical reflective confocal microscopic imaging system can be solved through the same method as the optical transfer function of the elliptical mirror in incoherent illumination.

Figure 7.8 is a schematic diagram of the self-convolution of a spherical shell. One spherical shell reverses during self-convolution. The focusing angle of any point P on the space intersecting line on the two spherical surfaces is calculated by reference to the deduction in section 7.1 as follows:

$$\cos \theta_{1,2} = \frac{|s|}{2}(1 \mp p \cos \beta).$$

(7.23)

Here, the expression of p is the same as the expression of p in the deduction of the three-dimensional optical transfer function of the elliptical mirror, as shown in formula (7.15), β represents the included angle between MO and PO, the coherent transfer function $c(l, s)$ can be expressed as the integral function of β, and the integral limit β_2 is

$$\beta_2 = \arcsin\left[\frac{1}{p}\left(1 - \frac{2\cos \alpha}{|s|}\right)\right]$$

(7.24)

where,

$$|s| \geqslant 2 \cos \gamma.$$

(7.25)

Especially, all parameters meet

$$2(|l|\sin \gamma - |s|\cos \gamma) \geqslant l^2 + s^2.$$

(7.26)

The intersecting line of two spherical shells is a complete circle, and then $\beta_2 = \pi/2$, so

$$\beta_2 = \begin{cases} \arcsin\left[\dfrac{1}{p}\left(1 - \dfrac{2\cos\alpha}{|s|}\right)\right] & l^2 + s^2 \leqslant 2(|l|\sin\gamma - |s|\cos\gamma),\ |s| \geqslant 2\cos\gamma \\ \pi/2 & 2(|l|\sin\gamma - |s|\cos\gamma) \leqslant l^2 + s^2 \leqslant 4,\ |s| \geqslant 2\cos\gamma \end{cases} \quad (7.27)$$

Here is the integral form that the coherent transfer function of the elliptical reflective confocal microscopic imaging system is expressed as the product of apodization factors $P(\theta)$.

$$C(l, s) = \frac{4}{\pi\sqrt{l^2 + s^2}} \int_0^{\beta_1} P_1(\theta_1)P_2(\theta_2)\, d\beta. \quad (7.28)$$

The zero-free region of the coherent transfer function $c(l,s)$ of the elliptical reflective confocal microscopic imaging system is a spherical cap

$$l^2 + s^2 = 4 \qquad s = \pm 2\cos\gamma. \quad (7.29)$$

The lateral spatial cutoff frequency is $2\sin\gamma$, and the axial spatial cutoff frequency is $[\pm 2\cos\gamma, \pm 2]$.

Figure 7.9 is a three-dimensional graph of the three-dimensional coherent transfer function. Obviously, under the condition of high numerical aperture, the coherent transfer function of the elliptical reflective confocal microscopic imaging system is greatly increased, and the frequency domain cutoff frequency is expanded. Therefore, it can be judged that the confocal system has the resolving power for high spatial frequency information.

Figure 7.10 shows the coherent transfer function of the confocal microscope consisting of lenses. Accordingly, under the condition of the same numerical

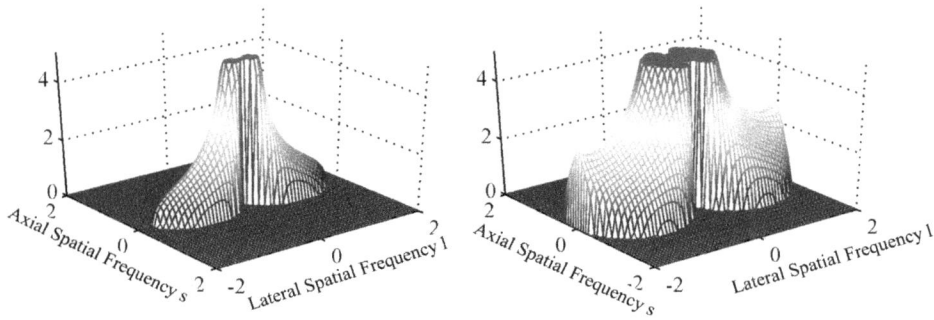

a) The maximum focusing angle is $\pi/3$.   b) The maximum focusing angle is $\pi/2$.

Figure 7.9. Three-dimensional coherent transfer function of the elliptical reflective confocal microscopic system.

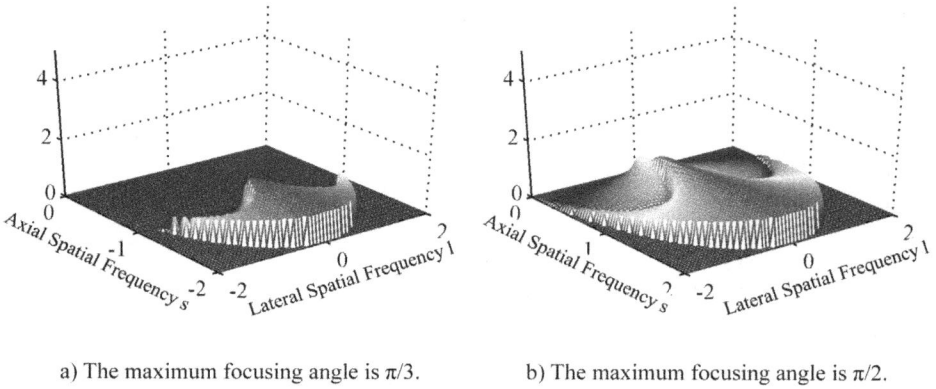

a) The maximum focusing angle is $\pi/3$.　　　b) The maximum focusing angle is $\pi/2$.

Figure 7.10. Three-dimensional coherent transfer function of a traditional lens confocal microscopic system (Sine condition).

aperture, the three-dimensional coherent transfer function of the elliptical mirror increases in all frequency bands compared with that of the lens, so the elliptical mirror has higher resolving power, while under the condition of a high numerical aperture, this advantage of the elliptical mirror is particularly prominent.

7.3.2 Two-dimensional transfer function of the elliptical reflective confocal microscopic imaging system

Prof. M Gu provides the relation between the three-dimensional transfer function and the two-dimensional lateral modulation transfer function (MTF) of the optical system [1] as follows:

$$c(l, s) = \int_{-\infty}^{\infty} c(l, u)\exp(-ius)\, du \qquad (7.30)$$

where u is the axial optical coordinate. It can be seen that the three-dimensional transfer function is the Fourier transform of MTF along the axial direction as we know it, so we can get the expression form of the inverse Fourier transform of the two-dimensional lateral transfer function of the optical system.

$$c(l, u) = \int_{-\infty}^{\infty} c(l, s)\exp(ius)\, ds. \qquad (7.31)$$

Especially, we are concerned with the resolving power of the optical system on a focal plane. Therefore, supposing $u = 0$, MTF on the focal plane is

$$c_{focal}(l) = \int_{-\infty}^{\infty} c(l, s)\, ds. \qquad (7.32)$$

Figure 7.11 shows MTF of the elliptical reflective confocal microscopic imaging systems under the condition of different numerical apertures, all the curves are

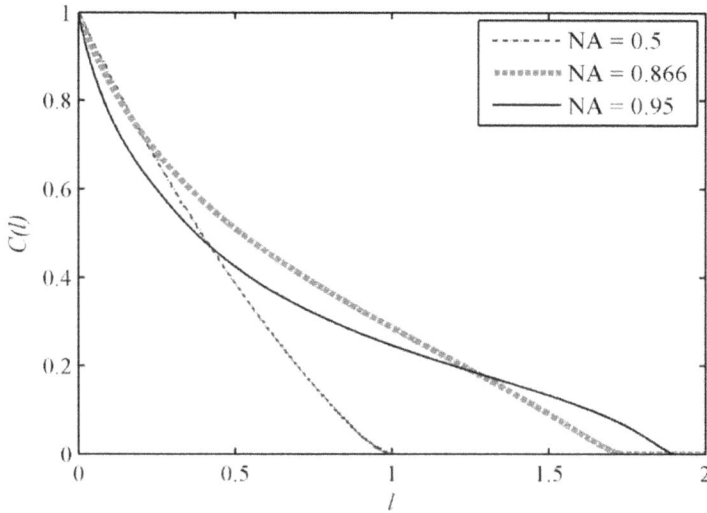

Figure 7.11. MTF curves of the confocal system consisting of elliptical mirrors with different numerical apertures.

normalized with the value at $l = 0$. It can be seen that with the increase of the numerical aperture, the spatial cutoff frequency of the optical system is expanded, the transfer function increases obviously at high-frequency band as a cost of decrease at low-frequency band, and especially when the maximum focusing angle of the elliptical mirror is $\pi/2$, the low-frequency band decreases severely.

Accordingly, the increase and decrease of MTF of the elliptical mirror are directly reflected in the change of imaging contrast. The cutoff frequency is expanded under the condition of high numerical aperture, but it only indicates that subject to the current detection technology, the optical system can image the added high-frequency information. The useful information will be overwhelmed by noise due to low contrast, and thus cannot be detected. Meanwhile, the whole imaging resolution may not be enhanced due to apparent decline in the contrast of the low-frequency component. Therefore, we shall search for the optical system with the maximum MTF at this frequency band according to the frequency characteristics of the information to be observed instead of evaluating the resolving power of the imaging system only by the description of the optical transfer function. Some relevant scholars also propose to reduce the noise of the receiving end by reducing the temperature and dimension of the detector [3].

We compare the MTF curve of the traditional confocal system with that of the elliptical reflective confocal system, as shown in figure 7.12.

The analysis result indicates when the elliptical mirror and the focusing lens have the same numerical aperture, the MTFs of the elliptical reflective confocal microscopic system and the traditional confocal system basically coincide under the condition of small numerical aperture (NA = 0.5), which indicates that they have the same lateral resolution, as shown in figure 7.12(a). But when the numerical aperture

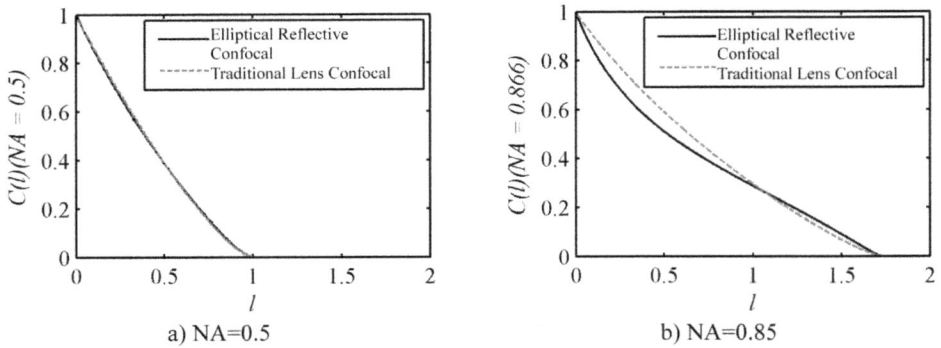

Figure 7.12. Comparison of MTF curves of elliptical reflective confocal system and traditional confocal system.

is 0.85, as shown in figure 7.12(b), the MTF of the elliptical reflective confocal microscopic system decreases at a low-frequency band and increases at a high-frequency band compared with the lens system, which is more obvious with the further increase of the numerical aperture.

Therefore, the essence of the elliptical reflective confocal imaging system for having a high theoretical resolution is the expansion of the frequency domain cutoff frequency and the increase of the transfer function at the high-frequency band, so that the system has the detectivity for high frequency information, and thus can distinguish the high frequency details of the sample, realizing the theoretical measurement of high resolution.

7.4 Summary

In this chapter, we theoretically analyze and deduce the point spread function, the coherent transfer function in the coherent illumination and the optical transfer function in the incoherent illumination of an elliptical mirror, and we find the elliptical mirror has a better performance than the lens in the high-frequency information distribution. Specifically, we use the three-dimensional transfer function theory to study the imaging characteristics of the elliptical reflective confocal microscopic system and find the high-frequency information distribution of the objective enhances due to the effect of the elliptical mirror. Given that the elliptical mirror will have a bright future in imaging because of its characteristics as a reflective component mentioned in the previous chapters, we believe that the analysis of transfer function will be an important reference in the research of imaging.

References

[1] Gu M 1996 *Principles of Three-Dimensional Imaging in Confocal Microscopes* (Singapore: World Scientific) pp 273–311
[2] Gu M 2000 *Advanced Optical Imaging Theory* (Berlin: Springer) pp 71–109 143–198.
[3] Pawley J B 2006 *Handbook of Biological Confocal Microscopy* 3rd edn (Berlin: Springer) pp 20–43

IOP Publishing

Elliptical Mirrors
Applications in microscopy
Jian Liu

Chapter 8

Design and application of an aspherical mirror

Yong Li, Jian Liu, Chao Wang and Jiubin Tan

8.1 Introduction

In the previous chapter, we analyzed some optical properties and imaging characteristics of an aspherical reflector (ellipsoid mirror). The aspherical reflector (ellipsoid mirror) has a number of wonderful properties, such as the second-order aspherical mirror increases the y-axis resolution by 12.07% compared with the aplanatic lens (chapter 2), the lateral and axial FWHMs of the elliptical mirror are 90% of the lens (chapter 5), free of chromatic aberration (chapter 1) and so on. The reflective imaging system is widely used in such fields requiring a large aperture scale as astronomical telescope and space telescope, as well as the special microscopy field that has a strict requirement for chromatic aberration. Fortunately, due to the development of technology, we do not need to worry about the accuracy of aspheric surface machining. However, how do we design it? How do we build an aspherical reflective imaging system? We will give some discussions and share the experience in this chapter.

Since Karl Schwarzschild put forward the reflective objective in 1905, great progress has been made in the theory and application of the reflective objective, especially, mature design model and perfect theoretical system have been formed for spherical reflective objective (Schwarzschild objective) [1], and an optical system designer can design the spherical reflective objective which he needs according to the design model. The design method of the Schwarzschild objective is simple but has deficiency in principle, that is the obscuration is not taken into consideration. Compared with a spherical system, an aspherical system has obvious advantages in reducing the number of lenses, improving image quality, optimizing system structure and design freedom. However, the design model of an aspherical system is more complex, so the research on aspherical reflective objective is still less, and there are no mature model and theoretical system for optical system designers to refer to.

The current design model of the aspherical reflective objective mainly makes reference to the design model of the aspherical reflective objective in a polar

doi:10.1088/978-0-7503-1629-3ch8
8-1

coordinate system put forward by Head [2] in 1957, but in the design model of the aspherical reflective objective in polar coordinate system, the curvature radius of the aspheric surface's vertex is coupled with the aspherical coefficient, so the analytical solutions to the radius and the aspherical coefficient cannot be directly obtained, and the actual engineering application capacity of this model is not high [3].

Compared with the spherical reflective objective, the aspherical reflective objective has the advantages of high design freedom, flexible structure and the like, but the increase in the design parameters results in a too complex design model of polar coordinates, therefore, it cannot be used in practical design and analysis. Therefore, obtaining a simple and effective design model of aspherical reflective objective is one of the important scientific problems in the current design theories of the aspherical reflective objective. For the coaxial reflective microscopy imaging, the obscuration caused by the secondary mirror of the reflective objective is one of the important factors that affect the image quality. The secondary mirror blocks some part of light from the object plane, and only the light at the edge contributes to the imaging. The existence of large obscuration not only reduces the brightness of the image, but also lowers the image quality. The problem of large obscuration caused by the secondary mirror of the reflective objective is one of the major reasons that restrain the development of the reflective objective. The existing reflective objective model can be used to solve the structural parameters of the reflective objective, but the obscuration proportion, which is the key parameter, cannot be effectively controlled. Therefore, establishing a design model of reflective objective structural parameters with a constraint on the obscuration ratio is an urgent scientific problem in the design theories of the reflective objective. This chapter mainly introduces the analytical design model and obscuration constraint model of an aspherical reflective objective in a rectangular coordinate system based on the Taylor series, solves the problem that the curvature radius of the aspheric surface's vertex and the aspherical coefficient cannot be decoupled and the problem that the obscuration caused by the secondary mirror is large in the design of the reflective objective, further improves the design theory system of reflective objective, provides a theoretical basis for the development of aspherical reflective microscopy imaging, and uses this design model to specifically develop a microscopical instrument based on the reflective aspherical objective with extra-long working distance, wide spectrum, apochromatism and high numerical aperture.

8.2 Basic knowledge

In order to ensure the fluency of analysis and the completeness of theory in the text below, this section briefly introduces the basic knowledge of the aspherical surface and Taylor series involved in some theoretical analyses.

8.2.1 Mathematical representation of aspherical surface

The so-called aspherical surface means all types of surface except spherical surface and plane surface, and can be roughly divided into a non-rotational symmetric aspherical surface, a rotational symmetric aspherical surface, an array surface and aspherical surface without a center of symmetry. Compared with the other three types of aspherical surface, the rotational symmetric aspherical surface is relatively simple in design, processing and detection, and is therefore the most widely used aspherical surface in conventional optical system. The aspherical surface to be discussed in this chapter refers to rotational symmetric aspherical surface.

A rotational symmetric aspherical surface is usually a curved surface formed by rotation of a quadratic curve or high-order curve around its axis of symmetry. For example, the corresponding curves of a hyperboloid and paraboloid widely used in an astronomical telescope are respectively a hyperbolic curve and a parabola. If it is assumed that, in a three-dimensional rectangular coordinate system, the rotation axis is the z-axis (the rotation axis is also the optical axis), the origin of coordinates is the aspheric surface's vertex, and the distance from any point on the aspherical surface to the optical axis is r ($r^2 = x^2 + y^2$), then the expression of a rotational symmetric aspherical surface is as follows:

$$z = \frac{cr^2}{1 + \sqrt{1 - (1 + k)c^2 r^2}} + \beta_1 r^1 + \beta_3 r^3 + \beta_5 r^5 + \cdots. \tag{8.1}$$

In formula (8.1), c is the curvature at the aspheric surface's vertex (i.e. the reciprocal of radius, $c = 1/R$, where R is the curvature radius of the vertex); k is the quadric surface coefficient or the conic coefficient ($k = -e^2$, where e is the excentricity or eccentricity); z represents the corresponding vertical distance or is called the vector height of the curved surface; and β_1, β_3 and β_5 are called the high-order coefficients. The first term represents the benchmark quadric surface, and the following terms are the high-order terms of the aspherical surface. The high-order terms represent the surface height deviation from the quadric surface, i.e. the aspherical surface is obtained by superimposing some small surface deformation caused by the high-order terms on the quadric surface to achieve the purpose of correcting aberration. The high-order coefficients may or may not be zero. Compared with the spherical surface, the aspherical surface is added with multiple optimizable variables, which can control the direction of the light passing through the surface. Therefore, the aspherical surface has advantages over the spherical surface in design. In some systems, an aspherical surface can be used to replace several spherical structures to achieve better effects or simplify the complex optical system structure.

When the high-order coefficients are all zero, only a quadric surface (also called a conic surface) remains. The quadric surface is the simplest kind of aspherical surface, as shown in figure 8.1, and the expression of the quadric surface is as follows [6]:

$$z = \frac{cr^2}{1 + \sqrt{1 - (1 + k)c^2 r^2}}. \tag{8.2}$$

This expression represents a standard quadric surface, also called a conic surface. The value of the conic coefficient k determines the shape of the quadric surface: when $k < -1$, the expression represents a hyperboloid; when $k = -1$, the expression represents a paraboloid; when $-1 < k < 0$, the expression represents an ellipsoid (the major axis overlaps with the optical axis); when $k = 0$, the expression represents a

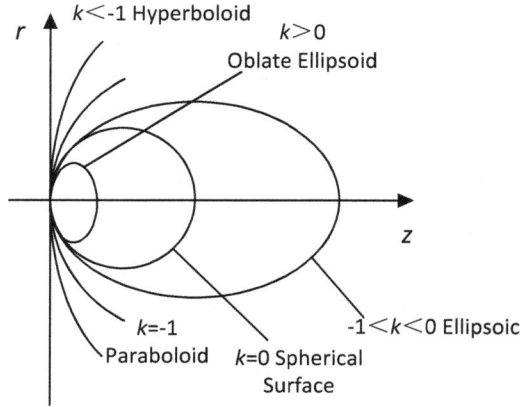

Figure 8.1. Conic surfaces.

spherical surface; and when $k > 0$, the expression represents an oblate ellipsoid (the minor axis overlaps with the optical axis). If it is assumed that a and b are respectively the major axis and the minor axis of the ellipsoid, then:

$$k = -e^2 = -\left[\frac{a^2 - b^2}{a^2}\right]. \tag{8.3}$$

Expression (8.1) is just a representation of the aspherical surface. Since the high-order terms are all the odd-order terms of r, it is called the odd-order aspherical surface. Correspondingly, the even-order aspherical surface has all the high-order terms be the even-order terms of r, and its expression is as follows:

$$z = \frac{cr^2}{1 + \sqrt{1 - (1 + k)c^2r^2}} + \alpha_1 r^2 + \alpha_2 r^4 + \alpha_3 r^6 + \alpha_4 r^8 + \cdots. \tag{8.4}$$

There is no essential difference between the two expressions, and in practical application, the expression of the even-order aspherical surface is used more often. The formulas discussed above are the mathematical expressions of aspherical surface in the three-dimensional rectangular coordinate system. However, in some applications, the aspherical surface needs to be represented in polar coordinate system. In this case, the expressions in rectangular coordinate system can be transformed into ones in a polar coordinate system by the coordinate transformation relation. The coordinates of the aspherical surface in polar coordinate system are composed of three parameters (ρ, φ, θ), and their value ranges are:

$$\begin{aligned}
\rho &\in [0, +\infty], \\
\phi &\in [0, 2\pi], \\
\theta &\in [0, 2\pi].
\end{aligned} \tag{8.5}$$

The principle for transforming the coordinates of the aspherical surface in polar coordinate system into the coordinates in rectangular coordinate system is shown in

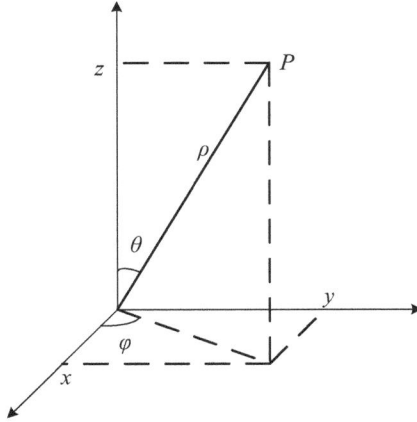

Figure 8.2. Transformation between Cartesian and polar coordinates.

figure 8.2. For a point P in the space, the transformational relation between the coordinate representation (ρ, φ, θ) in polar coordinate system and the coordinate representation (x, y, z) in rectangular coordinate system is as follows:

$$
\begin{aligned}
x &= \rho \sin \theta \cos \phi, \\
y &= \rho \sin \theta \sin \phi, \\
z &= \rho \cos \theta.
\end{aligned}
\qquad (8.6)
$$

Conversely, the coordinates of the aspherical surface in rectangular coordinate system can also be transformed into coordinates in polar coordinate system. For a point P in the space, the transformational relation between the coordinate representation (x, y, z) in rectangular coordinate system and the coordinate representation (ρ, φ, θ) in polar coordinate system is as follows:

$$
\begin{aligned}
\rho &= \sqrt{x^2 + y^2 + z^2}, \\
\phi &= \arctan \frac{y}{x}, \\
\theta &= \arccos \frac{z}{\rho}.
\end{aligned}
\qquad (8.7)
$$

The three-dimensional coordinates in polar coordinate system and the three-dimensional coordinates in rectangular coordinate system can be transformed conveniently by formulas (8.6) and (8.7). The transformational relation between two-dimensional coordinate systems is consistent with the transformational relation between three-dimensional coordinate systems, and the only difference is the lack of z and θ.

8.2.2 Taylor series

Taylor's formula is a common mathematical tool, which uses the information of a function at a certain point to describe the value near this point. If the function is smooth enough, and in the condition that all order derivatives of the function at a certain point are known, Taylor's formula can be used to construct a polynomial to approximately represent the value of the function in the neighboring domain of this point.

If the function $f(x)$ has a derivative up to the nth order at point x_0, the Taylor's formula of $f(x)$ at point x_0 with a Peano remainder term is as follows:

$$f(x) = f(x_0) + \frac{f'(x_0)}{1!}(x - x_0) + \frac{f^{(2)}(x_0)}{2!}(x - x_0)^2 + \cdots$$
$$+ \frac{f^{(n)}(x_0)}{n!}(x - x_0)^n + R_n(x), \tag{8.8}$$

where, $R_n(x)$ is the Peano remainder term of $f(x)$:

$$R_n(x) = o[(x - x_0)^n]. \tag{8.9}$$

One of the great significances of Taylor series is that it can be used to approximately calculate the value of a function. The value of $f(x)$ near x_0 can be approximated by all order derivatives of this function, and the higher the order used, the closer the value of the function is.

When it is necessary to analyze and compare several different forms of functions, direct calculation may not be possible, but if several functions are expanded at the same point in the form of Taylor series expansion, the coefficients of each term can be compared in the form of a uniform power series. The higher the order of the series is, the more the term tends to be infinitesimal, and therefore the smaller the contribution is to the whole. In general, several functions can be accurately judged by comparing the first few terms of the Taylor series.

In order to simplify the calculation, in many cases, the function $f(x)$ is expanded at zero. Correspondingly, in Taylor's formula, the series for $x_0 = 0$ is as follows:

$$f(x) = \sum_{n=0}^{\infty} \frac{f^{(n)}(0)}{n!} x^n + o(x^n), \tag{8.10}$$

where this formula is called the Maclaurin formula, which is a frequently used special case of Taylor's formula.

8.3 Design of reflective objective

Compared with a refractor, a mirror has the great advantage of no chromatic aberration, which makes it have an incomparable advantage in the field of wide spectrum imaging. However, the mirror is not free of aberration, and instead, it is only free of aberration in some cases, for example at the curvature vertex of plane mirror and spherical mirror. The study on the aberration of spherical and aspherical mirrors is detailed in the book *Reflective Optics* written by D Korsch [6]. This chapter mainly introduces the design model of reflective objective, and does not introduce the aberration. For the convenience of understanding our design model of aspherical reflective objective, we will introduce in detail the design model of aspherical reflective objective put forward by Head in this section.

8.3.1 Head design model

The aspherical optics can better reduce the number of lenses, improve image quality and optimize system structure than the spherical optics. Therefore, with the continuous improvement in ultra-precision processing technology and optical detection technology, the aspherical surface has been widely used in the system that requires a high level of image quality, and the application of aspherical surfaces is an irresistible trend of development of the optical system.

Since the early research of Karl Schwarzschild, all of the traditional research methods have been based on the approximation of Taylor series expansion of trigonometric function, and have involved the spherical mirror. In 1957, Head proposed designing an aplanatism objective without approximation, and deduced an accurate aspherical analytical model under the Abbe sine condition and the axis astigmatism conditions in polar coordinate system.

Head objective model is to deduce a polar coordinate model of aspherical reflective objective with the starting point of correcting the spherical aberration of the points on axis. The structure of aspherical reflective objective [2] is shown in figure 8.3, where OPQI is a ray from object point O to image point I, and P and Q are respectively the intersection points of the ray with the primary mirror and the secondary mirror. The polar coordinate expression of the primary mirror is expressed by polar coordinates (ρ, θ), where O is the pole, and $-\pi/2 < \theta < \pi/2$; the polar coordinate expression of the secondary mirror is expressed by polar coordinates (r, u), where I is the pole. The incident angle of the ray relative to the normal line of the primary mirror is i, and the distance of PQ is l. ρ, l and r are projected onto the optical axis and the corresponding distances are respectively ρ_0, l_0 and r_0.

Head pointed out that this reflective objective need to satisfy formulas (8.11)–(8.15) in the condition of eliminating spherical aberration and coma, and the expressions of the polar coordinates of the primary mirror and the secondary mirror are obtained by simplifying the five formulas.

The optical path difference of the on-axis ray is the same as that of the off-axis ray, and shall satisfy:

$$\rho - l + r = \rho_0 - l_0 + r_0. \tag{8.11}$$

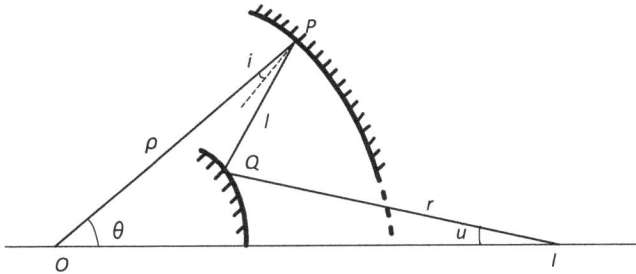

Figure 8.3. Principle of the Head objective.

When eliminating spherical aberration and coma, the system shall meet the sine condition:

$$\sin \theta = m \sin u. \tag{8.12}$$

When the light is incident on the primary mirror, the reflection angle shall satisfy:

$$\frac{1}{\rho}\frac{d\rho}{d\theta} = - \tan i. \tag{8.13}$$

The projection of PQ parallel to the optical axis and that perpendicular to the optical axis shall satisfy:

$$l \cos(2i + \theta) = \rho \cos \theta + r \cos u - (\rho_0 + r_0 - l_0). \tag{8.14}$$

$$l \sin(2i + \theta) = \rho \sin \theta - r \sin u. \tag{8.15}$$

The way to simplify these formulas is to: eliminate the variables r, l and u of formulas (8.11), (8.12), (8.14) and (8.15), and then substitute the expression of $\tan i$ into formula (8.13) to obtain a differential equation of ρ and θ. This equation is the polar coordinate expression of the primary mirror.

Head used the following transformation during the derivation:

$$\gamma = \cos \theta + \sqrt{m^2 - \sin^2 \theta}, \tag{8.16}$$

that is:

$$\cos \theta = \frac{\gamma^2 - m^2 + 1}{2\gamma}, \tag{8.17}$$

where, m represents the magnification of the system; θ is the angle between the object space emergent ray and the optical axis.

Ultimately, Head obtained two expressions of aspherical mirror in polar coordinates, of which the expression of aspherical equation of the primary mirror is as follows:

$$\frac{l_0}{\rho} = \frac{1 + \kappa}{2\kappa} + \frac{1 - \kappa}{2\kappa}\cos \theta + \left[\frac{l_0}{\rho_0} - \frac{1}{\kappa}\right]\left[\frac{\gamma}{1 + m}\right]^{-1}\left[\frac{\gamma - (1 - m)}{2m}\right]^{\alpha}$$

$$\times \left[\frac{\gamma - (m - 1)}{2}\right]^{\beta}\left[\frac{\kappa + 1}{2(m + 1)}\gamma - \frac{\kappa - 1}{2}\right]^{2-\alpha-\beta}, \tag{8.18}$$

where, $\kappa = (\rho_0 + r_0)/l_0$, $\alpha = m\kappa/(m\kappa - 1)$, and $\beta = m/(m - \kappa)$.

The equation of the secondary mirror can be obtained by the reversibility of the system:

$$\frac{l_0}{r} = \frac{1+\kappa}{2\kappa} + \frac{1-\kappa}{2\kappa}\cos u + \left[\frac{l_0}{r_0} - \frac{1}{\kappa}\right]\left[\frac{\delta}{1+M}\right]^{-1}\left[\frac{\delta - (1-M)}{2M}\right]^{\alpha'}$$

$$\times \left[\frac{\delta - (M-1)}{2}\right]^{\beta'}\left[\frac{\kappa+1}{2(M+1)}\delta - \frac{\kappa-1}{2}\right]^{2-\alpha'-\beta'},$$

(8.19)

where $\alpha' = M\kappa/(M\kappa-1)$, $\beta' = M/(M-\kappa)$, $M = 1/m$, and $\delta = M\gamma = \cos u + \sqrt{M^2 - \sin^2 u}$.

Formulas (8.18) and (8.19) are the polar coordinate expressions of models of the primary mirror and the secondary mirror with ideal surface shapes in the condition of eliminating spherical aberration and coma. As the optical path has a symmetrical structure about the optical axis, the expressions represent rotational symmetric aspherical surface.

Although Head obtained the polar coordinate expressions of models of the primary mirror and the secondary mirror, the parameters were coupled together, therefore it was unable to separate the curvature radius of the aspheric surface's vertex, the conic coefficient and the high-order terms of the aspherical surface, unable to obtain specific parameters, and still difficult to play a guiding role in actual engineering design.

It is worth mentioning that the Schwarzschild objective is made of two spherical mirrors concentric to each other, the previous research on the Schwarzschild objective mainly focused on the imaging characteristics of spherical reflection, so did the calculation. However, the only one variable parameter in the spherical surface design is the radius, and it is difficult to satisfactorily improve the image resolution and lower the obscuration ratio of the system simultaneously by the design of spherical radius parameters, therefore, the introduction of quadric conic surface and even the high-order aspherical surface into the system is taken into consideration. Because of the introduction of the aspherical coefficient, the design freedom can be increased to meet the requirements of imaging characteristics and shading ratio at the same time.

8.3.2 Aperture diaphragm and field diaphragm

In addition to the image magnification and the object-image conjugate relation, the imaging range (also called the field of view) and the imaging beam aperture angle shall also be considered in optical system design. The imaging range determines the horizontal dimension of the field of view, and the imaging aperture angle determines the horizontal resolution of the system. In an optical system, the diaphragm limiting the imaging range is called a field diaphragm, and the diaphragm limiting the imaging beam aperture angle is called an aperture diaphragm.

The aperture diaphragm limits the imaging beam of the system. When the aperture diaphragm images in the object space through the optical system in front,

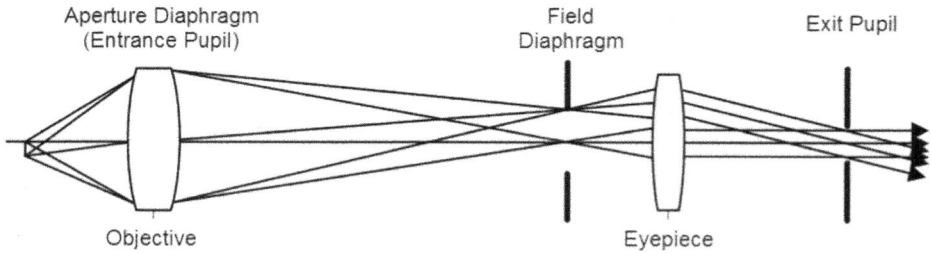

Figure 8.4. Diaphragm in a microscope.

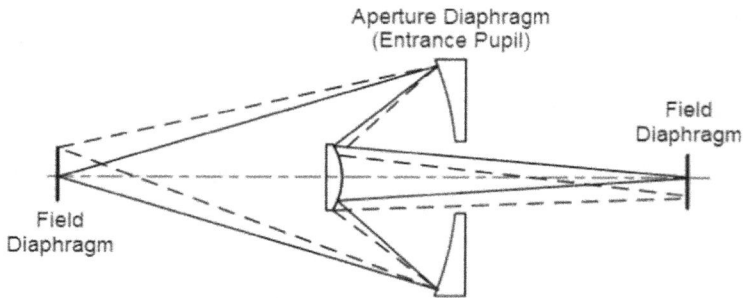

Figure 8.5. Diaphragm in the reflective objective.

the corresponding image determining the aperture size of the incident beam of the system is called the entrance pupil. When the aperture diaphragm images in the image space through the optical system at back, the corresponding image determining the aperture size of the emergent beam of the system is called the exit pupil. The beam limiting in a conventional microscope system is shown in figure 8.4. The frame of the objective is the aperture diaphragm and also the entrance pupil; and the image of the frame formed through the eyepiece is the exit pupil. For the visual microscope, the position of the exit pupil usually coincides with the position of the pupil of human eye. The field diaphragm is located in the image plane of the image space of the objective, and the field of view can be controlled by adjusting the size of the field diaphragm.

For reflective objective, if no other diaphragms are added, the frame of the primary mirror can be used as the aperture diaphragm and also the entrance pupil, and the size of the frame limits the aperture size of the imaging beam. The field diaphragm is located in the image plane of the image space of the objective. If a CCD camera is placed in this position as an imaging element, the size of the photosensitive chip of the CCD camera limits the imaging range, therefore the CCD chip is the field diaphragm, and its beam limiting is shown in figure 8.5.

8.4 Decoupled model based on the Taylor series expansion

As the surface of the reflector that we discussed in this chapter is rotational symmetric aspherical surface, the expression of the primary mirror can be expressed by the rotational symmetric aspherical equation. The Taylor series expansion is made respectively to the expression of the primary mirror and the rotational symmetric aspherical equation, and the coefficients of the corresponding order in the expanded expressions shall be equal. Assume that the primary mirror is a quadric aspherical surface, the two equations can be established when expanded into the quadratic terms. Solve the curvature radius of the vertex and the conic coefficient. If the primary mirror is a high-order aspherical surface, further expansion can be carried out by Taylor series to solve the high-order aspherical coefficient. The same method can be used to solve the surface shape parameters of the secondary mirror.

8.4.1 Taylor series expansion of the Head polar coordinate model

The decoupled model is studied by taking aspherical reflective objective as an example. Firstly, Taylor series expansion is made to the primary mirror equation obtained by the Head objective model. For the convenience of discussing below, the formula (8.18) of the primary mirror equation of the aspherical reflective objective is simplified:

$$\frac{l_0}{\rho} = \frac{1+\kappa}{2\kappa} + \frac{1-\kappa}{2\kappa}\cos\theta + \left[\frac{l_0}{\rho_0} - \frac{1}{\kappa}\right]y_1 y_2 y_3 y_4, \tag{8.20}$$

where:

$$\begin{cases} y_1 = \left[\dfrac{\gamma}{1+m}\right]^{-1}, \\[2mm] y_2 = \left[\dfrac{\gamma-(1-m)}{2m}\right]^{\alpha}, \\[2mm] y_3 = \left[\dfrac{\gamma-(m-1)}{2}\right]^{\beta}, \\[2mm] y_4 = \left[\dfrac{\kappa+1}{2(m+1)}\gamma - \dfrac{\kappa-1}{2}\right]^{2-\alpha-\beta}. \end{cases} \tag{8.21}$$

For simple operation, the following conversion is made:

$$t = 1 - \cos\theta = 2\sin^2(\theta/2). \tag{8.22}$$

According to section 8.3.1, $-\pi/2 < \theta < \pi/2$, then $0 < t < 1$. When the included angle θ approaches 0, t approaches 0 at a speed quadratic faster than θ, therefore a smaller truncation error can be obtained after expansion of the curved surface function based on t, and the accuracy is higher.

According to formulas (8.16) and (8.22), we can obtain:

$$\gamma = \sqrt{t^2 - 2t + m^2} + 1 - t. \tag{8.23}$$

Solve the first and the second derivatives of formula (8.21) with respect to t respectively. When $t = 0$, the derivative values are:

$$
\begin{cases}
y_1'(0) = \dfrac{1}{m},\ y_2'(0) = -\dfrac{m+1}{2m^2}\alpha, \\[2mm]
y_3'(0) = -\dfrac{m+1}{2m}\beta,\ y_4'(0) = -\dfrac{2-\alpha-\beta}{2m}(\kappa+1), \\[2mm]
y_1''(0) = \dfrac{m+1}{m^3},\ y_2''(0) = \dfrac{\alpha(\alpha-1)(m+1)^2}{4m^4} + \dfrac{\alpha(m+1)(m-1)}{2m^4}, \\[2mm]
y_3''(0) = \dfrac{\beta(\beta-1)(m+1)^2}{4m^2} + \dfrac{\beta(m+1)(m-1)}{2m^3}, \\[2mm]
y_4''(0) = \dfrac{(2-\alpha-\beta)(1-\alpha-\beta)(\kappa+1)^2}{4m^2} \\[2mm]
\qquad\quad + \dfrac{(2-\alpha-\beta)(\kappa+1)(m-1)}{2m^3}.
\end{cases}
\tag{8.24}
$$

Use t as the variable, make Taylor series expansion to formula (8.20) according to formula (8.8), and expand the formula into t quadratic terms:

$$
\begin{aligned}
\frac{l_0}{\rho} = \frac{1}{\kappa} &- \frac{1-\kappa}{2\kappa}t + \left(\frac{l_0}{\rho_0} - \frac{1}{\kappa}\right)\Big\{1 + \big[y_1'(0) + y_2'(0) + y_3'(0) + y_4'(0)\big]t \\[2mm]
&+ \Big\{y_1'(0)\big[y_2'(0) + y_3'(0) + y_4'(0)\big] \\[2mm]
&+ y_2'(0)\big[y_3'(0) + y_4'(0)\big] + y_3'(0)y_4'(0) \\[2mm]
&+ \frac{1}{2}\big[y_1''(0) + y_2''(0) + y_3''(0)y_4''(0)\big]\Big\}t^2 + o(t^2)\Big\}
\end{aligned}
\tag{8.25}
$$

where $o(t^2)$ is the Peano remainder term. By ignoring the high-order terms of formula (8.25) and simplifying it, we can obtain:

$$
\begin{aligned}
\frac{l_0}{\rho} = \frac{l_0}{\rho_0} &+ \Big\{\left(\frac{l_0}{\rho_0} - \frac{1}{\kappa}\right)\big[y_1'(0) + y_2'(0) + y_3'(0) + y_4'(0)\big] - \frac{1-\kappa}{2\kappa}\Big\}t \\[2mm]
&+ \left(\frac{l_0}{\rho_0} - \frac{1}{\kappa}\right)\Big\{y_1'(0)\big[y_2'(0) + y_3'(0) + y_4'(0)\big] \\[2mm]
&+ y_2'(0)\big[y_3'(0) + y_4'(0)\big] + y_3'(0)y_4'(0) \\[2mm]
&+ \frac{1}{2}\big[y_1''(0) + y_2''(0) + y_3''(0) + y_4''(0)\big]\Big\}t^2.
\end{aligned}
\tag{8.26}
$$

The above formula is the curved surface equation of the primary mirror of the aspherical reflective objective, which is expanded into the quadratic terms (the expansion order is 2).

8.4.2 Taylor series expansion of a quadric surface

Quadric surface is the simplest aspherical surface, and the decoupled model is derived by taking quadric surface as an example. Expand the quadric aspherical surface by the Taylor series into formula (8.26), and the curved surface coordinate of the quadric surface in rectangular coordinate system is as follows:

$$r^2 + z^2 - 2R_1z - e^2z^2 = 0, \tag{8.27}$$

where, R_1 is the curvature radius of the vertex and is in reciprocal relation with the curvature of the vertex c, and $R_1 = 1/c$; e is the eccentricity, and its relation with conic coefficient k is $k = -e^2$; z-axis is the direction of the optical axis.

In order to be consistent with the coordinate system of formula (8.18), the quadric surface represented by formula (8.27) has to be shifted along the z-axis. The shift distance is ρ_0, thus the curved surface equation of the quadric surface becomes:

$$(z - \rho_0)^2 + r^2 - 2R_1(z - \rho_0) - e^2(z - \rho_0)^2 = 0. \tag{8.28}$$

After shift, transform formula (8.28) into the polar coordinate system, and the curved surface coordinate of the quadric surface in polar coordinate system is as follows:

$$\frac{1}{\rho} = \frac{\cos\theta}{\rho_0} - \frac{\sqrt{(\rho_0^2 - e^2\rho_0^2 - 2R_1\rho_0 + R_1^2)\cos^2\theta - (\rho_0^2 - e^2\rho_0^2 - 2R_1\rho_0)} - R_1\cos\theta}{\rho_0^2 - e^2\rho_0^2 - 2R_1\rho_0}. \tag{8.29}$$

As the formula is too complex, it is changed by substitution, and let

$$w = \rho_0^2 - e^2\rho_0^2 - 2R_1\rho_0, \tag{8.30}$$

that is,

$$e^2 = \frac{\rho_0^2 - w - 2R_1\rho_0}{\rho_0^2}. \tag{8.31}$$

By substituting formula (8.31) into formula (8.29), the formula can be simplified as:

$$\frac{1}{\rho} = \frac{\cos\theta}{\rho_0} - \frac{\sqrt{(w + R_1^2)\cos^2\theta - w} - R_1\cos\theta}{w}. \tag{8.32}$$

The above formula is the curved surface equation of general quadric surface in polar coordinate system.

Use formula (8.22) to substitute the θ terms in formula (8.32) with t terms, and the curved surface equation of the quadric surface after substitution is as follows:

$$\frac{1}{\rho} = \frac{1}{\rho_0} - \frac{1}{\rho_0}t - \frac{\sqrt{(w + R_1^2)(t^2 - 2t + 1)} - w - R_1 + R_1t}{w}. \tag{8.33}$$

Calculate the values of the first and the second derivatives of this function for $t = 0$ respectively:

$$\begin{cases} \dfrac{d}{dt}\left(\dfrac{1}{\rho}\right)\Bigg|_{t=0} = \dfrac{1}{R_1} - \dfrac{1}{\rho_0}, \\[4mm] \dfrac{d^2}{dt^2}\left(\dfrac{1}{\rho}\right)\Bigg|_{t=0} = \dfrac{w + R_1^2}{R_1^3}. \end{cases} \qquad (8.34)$$

Expand formula (8.33) based on Taylor's formula using t as the independent variable. The formula is expanded into quadratic terms with high-order terms being ignored, and then multiplied by l_0, so the quadric surface equation is changed into:

$$\frac{l_0}{\rho} = \frac{l_0}{\rho_0} + \left(\frac{l_0}{R_1} - \frac{l_0}{\rho_0}\right)t + \frac{w + R_1^2}{2R_1^3}l_0 t^2. \qquad (8.35)$$

8.4.3 Determination of reflective objective parameters

Formula (8.26) represents the expression of Taylor series expansion of the primary mirror in polar coordinate system, and the expansion order is 2 due to expansion into the quadratic terms; while formula (8.35) represents the expression of Taylor series expansion of the quadric surface equation containing radius of curvature and conic coefficient, and the expansion order is also 2. In order to make two curved surfaces equal, the coefficients of corresponding terms of t in both formulas are equal. Since there are two parameters to be determined, we start from the terms with the lowest order to establish two equations, while the zero-order terms are the coordinates at the vertex of the mirror. We have used this condition to obtain the curved surface equation of Taylor expansion of the quadric surface and the primary mirror, so the zero-order terms of the two functions are certainly equal. Therefore, starting with the first-order terms of t, we only consider that the first-order terms and the second-order terms of t are respectively equal.

$$\frac{l_0}{R_1} - \frac{l_0}{\rho_0} = \left(\frac{l_0}{\rho_0} - \frac{1}{\kappa}\right)\left[y_1'(0) + y_2'(0) + y_3'(0) + y_4'(0)\right] - \frac{1 - \kappa}{2\kappa}. \qquad (8.36)$$

$$\begin{aligned} \frac{w + R_1^2}{2R_1^3}l_0 = \left(\frac{l_0}{\rho_0} - \frac{1}{\kappa}\right)\bigg\{ &y_1'(0)\left[y_2'(0) + y_3'(0) + y_4'(0)\right] \\ &+ y_2'(0)\left[y_3'(0) + y_4'(0)\right] + y_3'(0)y_4'(0) \\ &+ \frac{1}{2}\left[y_1''(0) + y_2''(0) + y_3''(0) + y_4''(0)\right]\bigg\}. \end{aligned} \qquad (8.37)$$

By further simplifying formula (8.36), we can obtain:

$$R_1 = -\frac{2ml_0\rho_0}{ml_0 + m\rho_0 - r_0}. \tag{8.38}$$

Formula (8.38) is the expression of the radius of curvature at the vertex of the primary mirror of the aspherical reflective objective.

It is worth mentioning that when considering the imaging of an ideal dual reflection system, it is known from the geometrical optics theory that

$$\begin{cases} \dfrac{1}{\rho_0} + \dfrac{1}{l'} = \dfrac{2}{R_1}, \\[2mm] \dfrac{1}{r_0} - \dfrac{1}{l' - l_0} = \dfrac{2}{R_2}, \\[2mm] \dfrac{r_0}{l' - l_0} \cdot \dfrac{l'}{\rho_0} = m. \end{cases} \tag{8.39}$$

By formula (8.39), we can obtain exactly the same expression as formula (8.38), and therefore the same conclusion is obtained by two methods. This indicates that the Taylor expansion method only considering that the first-order terms are equal is consistent with the paraxial calculation of the ideal spherical surface system, and the correctness of such series expansion method is also proved from another side.

By substituting formula (8.38) into formula (8.37), we can finally obtain:

$$\begin{aligned} w = {} & \frac{2r_0 R_1^3(m+1)}{\rho_0(\rho_0 + r_0)} \left[\frac{\kappa^3(m+1)^3(m-1)^2}{8m^3(m\kappa - 1)^2(m-\kappa)^2} + \frac{\kappa(m+1)(m-1)}{4m^2(m\kappa - 1)(m-\kappa)} \right. \\ & \left. - \frac{\kappa(\kappa + 1)^2(m-1)(2-\alpha-\beta)}{8m^2(m\kappa - 1)(m-\kappa)} + \frac{(2-\alpha-\beta)(\kappa+1)(\alpha+m\beta-1)}{4m^3} \right. \\ & \left. - \frac{(m+1)(\alpha+m\beta)}{4m^4} + \frac{1}{2m^3} \right] - R_1^2. \end{aligned} \tag{8.40}$$

By substituting formula (8.40) into formula (8.31), the conic coefficient of the quadric surface can be calculated, and the final expression of the conic coefficient of the primary mirror is as follows:

$$\begin{aligned} k_1 = {} & -e^2 = -\frac{(R_1 + \rho_0)^2}{\rho_0^2} - \frac{r_0 R_1^3(m+1)}{m^2 \rho_0^3(\rho_0 + r_0)} \left[\frac{\kappa^3(m+1)^3(m-1)^2}{4m(m\kappa - 1)^2(m-\kappa)^2} \right. \\ & \left. - \frac{\kappa(m+1)(m-1)}{2(m\kappa - 1)(m-\kappa)} + \frac{\kappa(\kappa+1)^2(m-1)(2-\alpha-\beta)}{4(m\kappa - 1)(m-\kappa)} \right. \\ & \left. - \frac{(2-\alpha-\beta)(\kappa+1)(\alpha+m\beta-1)}{2m} + \frac{(m+1)(\alpha+m\beta)}{2m^2} - \frac{1}{m} \right]. \end{aligned} \tag{8.41}$$

By substituting formulas (8.38) and (8.41) into formula (8.27), the decoupled curved surface model of the primary mirror of the aspherical reflective objective in rectangular coordinate system is obtained.

Similarly, according to the reversibility of the system, calculation of the secondary mirror can be made in the same way: if the image point is used as the object point, and the object point is used as the image point, then the primary mirror and the secondary mirror are also exchanged; at this point, the calculation method is not changed, so the calculation results are the parameters of the secondary mirror of the original system. Detailed derivation will not be made herein, and the formulas for calculating the radius of curvature and the conic coefficient of the secondary mirror are directly given as:

$$R_2 = \frac{2l_0 r_0}{r_0 + l_0 - m\rho_0},$$ (8.42)

$$
\begin{aligned}
k_2 = -e_2^2 = & -\frac{(R_2 - r_0)^2}{r_0^2} + \frac{\rho_0 R_2^3 (M + 1)}{M^2 r_0^3 (r_0 + \rho_0)} \left[\frac{\kappa^3 (M + 1)^3 (M - 1)^2}{4M(M\kappa - 1)^2 (M - \kappa)^2} \right. \\
& - \frac{\kappa(M + 1)(M - 1)}{2(M\kappa - 1)(M - \kappa)} + \frac{\kappa(\kappa + 1)^2 (M - 1)(2 - \alpha' - \beta')}{4(M\kappa - 1)(M - \kappa)} \\
& \left. - \frac{(2 - \alpha' - \beta')(\kappa + 1)(\alpha' + M\beta' - 1)}{2M} + \frac{(M + 1)(\alpha' + M\beta')}{2M^2} - \frac{1}{M} \right],
\end{aligned}
$$ (8.43)

where, R_2 is the radius of curvature of the secondary mirror, k_2 is the conic coefficient of the secondary mirror, $\alpha' = M\kappa/(M\kappa - 1)$, $\beta' = M/(M - \kappa)$, and $M = 1/m$. By substituting formulas (8.42) and (8.43) into formula (8.27), the decoupled curved surface model of the secondary mirror of the aspherical reflective objective in a rectangular coordinate system is obtained.

8.4.4 Derivation and truncation error of a high-order aspherical surface parameter

All of the above derivations are based on quadric surface, and the curvature radius of the vertex and the conic coefficient of the aspherical surface are obtained. For high-order aspherical surfaces, the same method can be used, and the high-order aspherical coefficient can be obtained by expanding into higher-order terms based on Taylor series. When solving the analytical expression of the high-order coefficients of the aspherical surface, the following expression of high-order aspherical function can be used:

$$z = \frac{cr^2}{1 + \sqrt{1 - (1 + k)c^2 r^2}} + fr^2 + gr^4 + hr^6 + o(r^6),$$ (8.44)

where f, g and the like are the aspherical coefficients of the second-, fourth- or even higher-order terms. Further Taylor series expansion is made to the polar coordinate model expression and high-order aspherical expression (8.44) of the primary mirror or the secondary mirror, more terms are retained as required, and the high-order coefficients are obtained successively by establishing multiple equations.

The above description is the derivation process of the decoupled model of reflective objective structural parameters based on Taylor series expansion.

During the derivation of this model, a physical model of the aspherical reflective objective in rectangular coordinate system is established and the analytical expressions of the curvature radius of the vertex and conic coefficient of the primary mirror and the secondary mirror are obtained by the method of Taylor series expansion. This model has a simple structure and strong applicability, which further improves the design theory system of the reflective objective.

It should be noted that although any parameters (including object distance, image distance and magnification) can be substituted into this model in theory, these parameters should be selected within a reasonable range conforming to optical principles, and the object distance, image distance and magnification should be matched as far as possible.

When the method of Taylor series expansion is used, the effect of truncation error on the final results should be considered. As the method of Taylor series expansion is used during the derivation of decoupled model of aspherical reflective objective, it is inevitable to generate truncation error, but this truncation error is a high-order error and corresponds to the high-order terms in the surface shape of the aspherical surface. For the design of an aspherical reflective objective, the most important thing is to determine the initial configuration parameters, that is to determine the curvature radius of the vertex and conic coefficient of the primary mirror and the secondary mirror; while the high-order coefficients corresponding to the truncation error can be further deduced by the high-order aspherical parameter derivation method introduced above, or the high-order coefficients can be further optimized by being set as the optimization variables in the simulation software.

8.4.5 Effect of numerical aperture on a decoupled model

Numerical aperture is one of the important indexes of objective design, and the effect of numerical aperture is not considered during the derivation of the decoupled model. In the above decoupled model, we use the quadratic surface approximation in order to simplify the decoupled principle and use the second-order Taylor series approximation which leads to the introduction of truncation error. In addition, from section 8.4.3, the above decoupled model is established on the condition of small numerical aperture. Thus, when the numerical aperture is increased, the image quality of the aspherical reflective objective obtained by using the decoupled model will inevitably be changed, and its system modulation transfer function curve is no longer close to the modulation transfer function curve of diffraction limit. The numerical aperture range applicable to the decoupled model of aspherical reflective objective based on Taylor series expansion can theoretically be obtained by further derivation and analysis of the system aberration, but this process involves a lot of complicated calculations. This section uses the example verification method to approximate the numerical aperture range applicable to the model from a practical point of view.

The numerical aperture NA is set as a variable, and other structural parameters are set as constants to verify that the numerical aperture range is applicable to this method. When $\rho_0 = 150$ mm, $l_0 = 60$ mm, $r_0 = 100$ mm and $m = 4$, the height of

Table 8.1. Parameters of reflective objectives.

Type	Radius of curvature R (mm)	Distance (mm)	Conic coefficient	Half aperture (mm)
Object plane	Infinity	150	0	0.3
Primary mirror	−97.2973	−60	−0.0968	−
Secondary mirror	−27.2727	100	−0.8591	−
Image plane	Infinity	−	0	1.2

object space field of view FOV is 0, 0.1 mm, 0.2 mm and 0.3 mm, the wavelength λ is set as 480 nm, 550 nm, 660 nm and 850 nm, and the object numerical aperture NA is respectively 0.1, 0.2, 0.3, 0.4, 0.5 and 0.6, the structural parameters of the aspherical reflective objective obtained by the decoupled model are shown in table 8.1.

As shown in table 8.1, only one set of structural parameters is given, this is because that for the different numerical apertures, the same structural parameters will be obtained by using the decoupled model, due to the fact the numerical aperture is not considered in the above decoupled model. But in fact, due to the use of the quadratic surface approximation and the second order Taylor series approximation, this set of structural parameters cannot satisfy all numerical apertures. We use modulation transfer function curves of the designed reflective objective in the cases of different numerical apertures to describe this conclusion intuitively. The corresponding modulation transfer function curves are shown in figure 8.6, and subgraphs (a)–(f) are the modulation transfer function curves of the reflective objective respectively corresponding to the numerical aperture NA of 0.1, 0.2, 0.3, 0.4, 0.5 and 0.6. The horizontal axis represents the spatial frequency, and the vertical axis represents the contrast. When the contrast drops to 0, the corresponding frequency is the cutoff frequency of the current system. From figure 8.6, we can see intuitively that, the larger the NA is, the greater the cutoff frequency of the system is. It is known from figure 8.6 that when the numerical aperture is small (NA = 0.1 and NA = 0.2), the modulation transfer function curve corresponding to each field of view of the aspherical reflective objective obtained by the decoupled model coincides with the modulation transfer function curve of diffraction limit, indicating that the image quality is good; as the numerical aperture increases to 0.3, the modulation transfer function curve deviates from the modulation transfer function curve of diffraction limit for each field of view, but the deviation is still in acceptable range; when the numerical aperture continues to increase to 0.4, 0.5 and 0.6, the modulation transfer function curve corresponding to each field of view seriously deviates from the modulation transfer function curve of diffraction limit (the larger the numerical aperture is, the greater the deviation is), indicating that the system has an aberration which seriously affects the image quality.

On the basis of the above design, further optimization is made respectively using Zemax in the cases of NA = 0.3, NA = 0.4, NA = 0.5 and NA = 0.6. The optimization method is to set the high-order terms of the primary mirror and

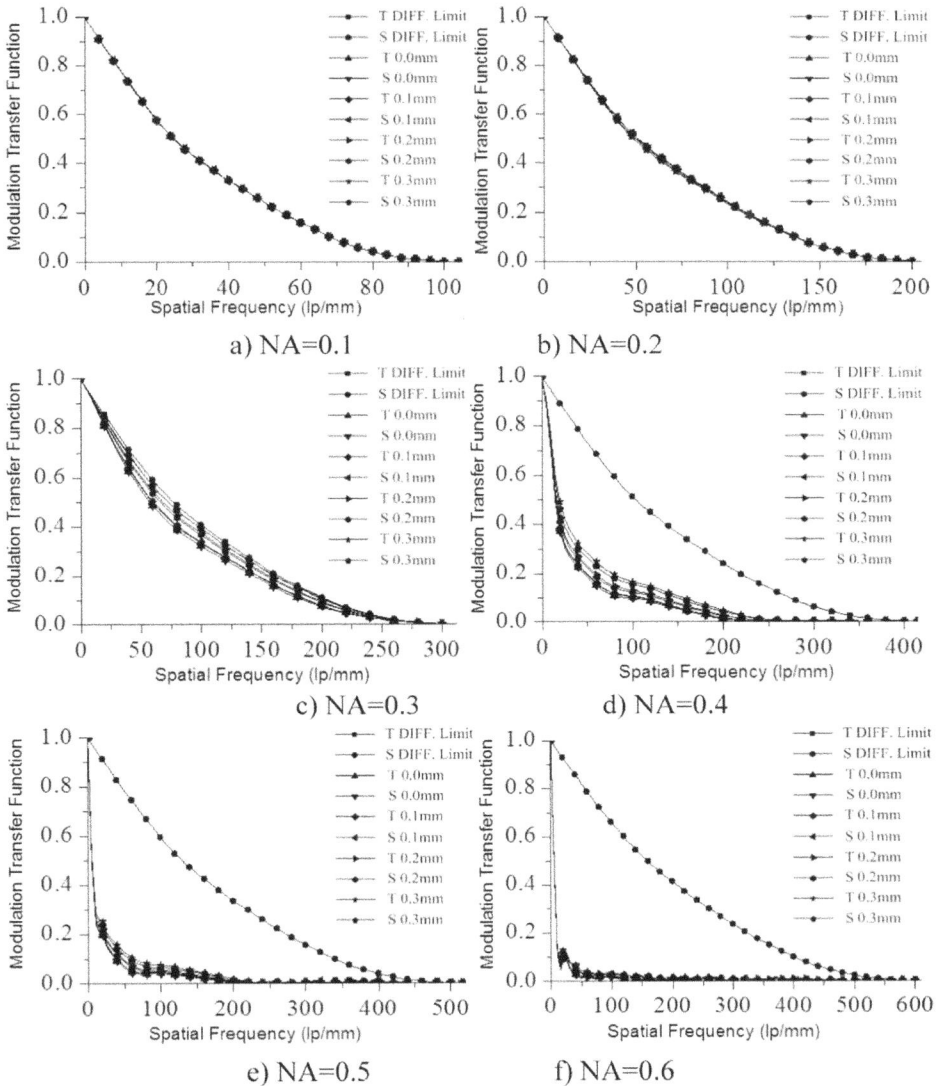

Figure 8.6. Modulation transfer function of reflective objectives with the change of NA.

the secondary mirror as optimization variables, with the highest order of 4 herein, and the optimization evaluation function adopts the default evaluation function of the system. The high-order coefficients of optimized aspherical surface are shown in table 8.2.

From table 8.2, in order to improve the imaging quality, the structures of the reflective objectives with different numerical apertures (NA = 0.3, 0.4, 0.5 and 0.6) are changed differently (different high-order coefficients) based on the same initial structure (table 8.1) to compensate the approximation error. The modulation

Table 8.2. High-order coefficients of an optimized reflective objective.

NA	Type	Second-order coefficient	Fourth-order coefficient
NA = 0.3	Primary mirror	0.000 004	$1.8442\ 62 \times 10^{-10}$
	Secondary mirror	0.000 150	$-2.833\ 54 \times 10^{-8}$
NA = 0.4	Primary mirror	0.000 005	$2.1874\ 85 \times 10^{-10}$
	Secondary mirror	0.000 192	$-6.459\ 84 \times 10^{-8}$
NA = 0.5	Primary mirror	0.000 007	$3.0287\ 37 \times 10^{-10}$
	Secondary mirror	0.000 263	$-1.132\ 11 \times 10^{-7}$
NA = 0.6	Primary mirror	0.000 010	$4.5734\ 52 \times 10^{-10}$
	Secondary mirror	0.000 364	$-1.782\ 28 \times 10^{-7}$

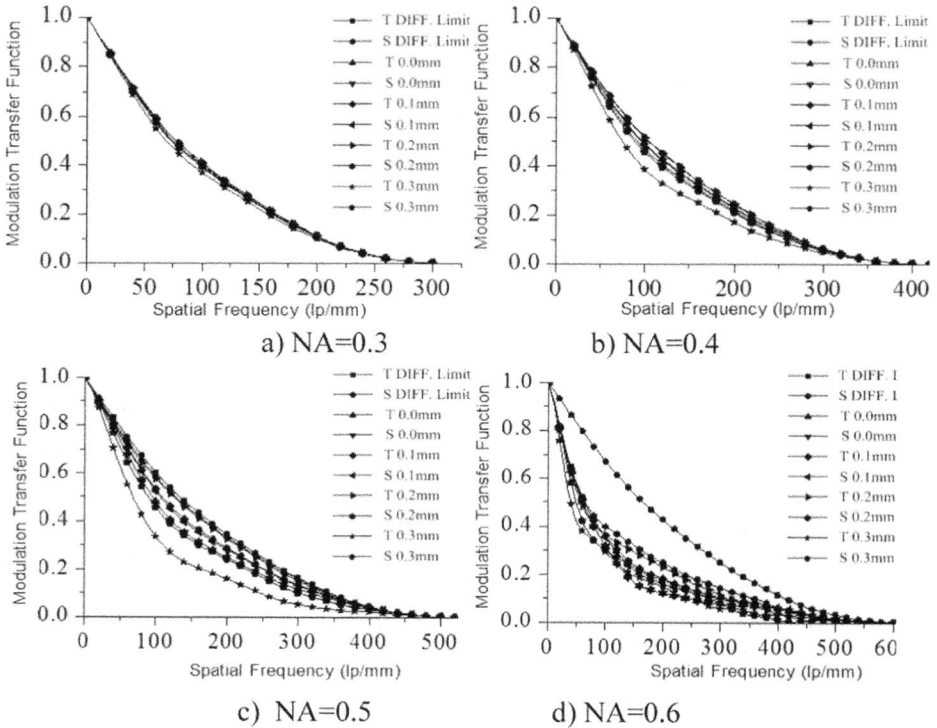

Figure 8.7. Modulation transfer function of optimized reflective objectives.

transfer function curves of the optimized system are shown in figure 8.7, and subgraphs (a)–(d) are the modulation transfer function curves of the optimized reflective objective respectively corresponding to the numerical aperture NA of 0.3, 0.4, 0.5 and 0.6. From subgraphs (a) and (b), we can see that the modulation transfer function curves of the system are close to the modulation transfer function curve of diffraction limit, especially the modulation transfer function curve corresponding to the field of view at the center, indicating that the system image quality is improved obviously after optimization of the high-order coefficient. With the increase of the numerical aperture, the modulation transfer function curve deviates gradually from the modulation transfer function curve of diffraction limit as shown in subgraphs (c) and (d). In subgraph (c), we can obviously see that, compared with the modulation transfer function curves corresponding to the field of view at the edge, the modulation transfer function curves corresponding to the field of view at the center are closer to the modulation transfer function curve of diffraction limit, indicating that the image quality of the field of view at the center is better than that of the field of view at the edge, because the starting points of the derivation of the Head reflective objective model and the decoupled model are both to correct the spherical aberration of the points on axis. When NA = 0.6 as shown in subgraph (d), the modulation transfer function curve of each field of view of the objective still has an obvious deviation from the modulation transfer function curve of diffraction limit, indicating that the image quality is poor.

The above example analysis shows that the decoupled model of reflective objective parameters based on Taylor series expansion is applicable to the design of reflective objective with small numerical aperture (NA < 0.3), and the initial configuration of which the image quality is close to the diffraction limit can be obtained directly by the decoupled model when reasonable parameters are selected; for the design of reflective objective with medium numerical aperture (NA < 0.5), ideal results can be obtained by further optimizing the high-order terms based on the decoupled model; when the numerical aperture is further increased (NA > 0.5), this model is no longer suitable.

8.5 Design method based on an obscuration constraint

8.5.1 Analysis of the obscuration effect on a reflective objective

The greatest disadvantage of the reflective objective is the obscuration of the secondary mirror to the imaging beam. Obscuration not only reduces the imaging energy of the system, but also affects the image quality of the system. The obscuration model of reflective objective is shown in figure 8.8. The object is located on the secondary mirror side and the corresponding image is located on the primary mirror side. When the imaging beam passes through the system, the secondary mirror obscures the beam at the center, and only the beam at the edge is involved in imaging, so this results in the reduction of information carrying capacity of the optical system in low and medium frequency bands. In the case that the object numerical aperture is constant, the larger the distance between the object and the secondary mirror is, the smaller the obscuration area is.

The concept of obscuration ratio is introduced here, which is the ratio of the obscuration area of the secondary mirror to the area of the imaging beam. The higher the obscuration ratio, the larger the obscuration area and the worse the image quality. The obscuration ratio can be calculated by the ratio of the area occupied by

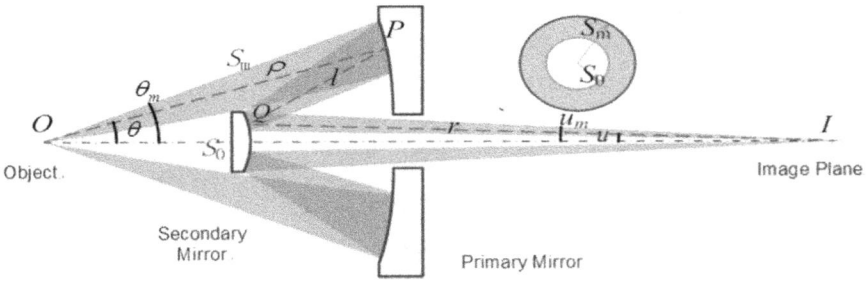

Figure 8.8. Obscuration model of the reflective objective [5].

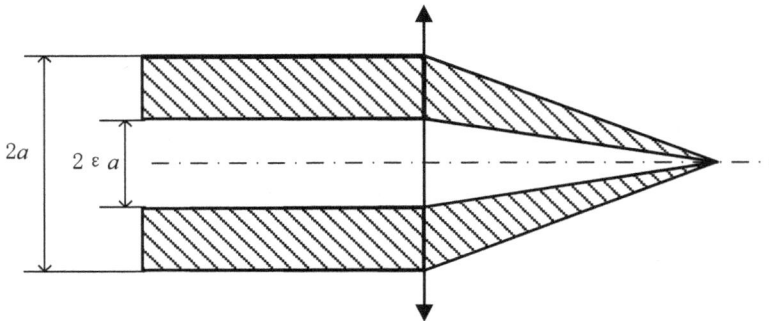

Figure 8.9. Simplified model of obscuration.

the secondary mirror in the plane where it is located to the total area of the light received by the plane. The calculation formula is as follows:

$$OR = \frac{S_0}{S_m}, \tag{8.45}$$

where, OR is the obscuration ratio; S_0 is the area obscured by the secondary mirror in the radial direction, and S_m is the total area of the light received by the plane where the secondary mirror is located.

The obscuration model of reflective objective is simplified as the model shown in figure 8.9. An ideal parallel light beam passes through an ideal lens and is focused. There is a central obscuration on the front surface of the lens, and the effect of the obscuration on system imaging is analyzed by calculating the modulation function of the system at different obscuration ratios.

For lens without obscuration, when the light beam is parallel to the axis, the amplitude distribution on the back focal plane of the lens can be expressed as follows:

$$U_3(r_3) = \frac{iU_0}{\lambda f} \exp(-ikf)\exp\left(-\frac{i\pi r_3^2}{\lambda f}\right)\int_0^\infty P(r_2)J_0\left(\frac{2\pi r_2 r_3}{\lambda f}\right)2r_2 dr_2, \tag{8.46}$$

where, P is the pupil function; r_2 is the lateral coordinate of the plane where the aperture is located; r_3 is the lateral coordinate of the focal plane; f is the focal distance; k is the wave number; and U_0 is the amplitude of the incident light. In the calculation below, let $U_0 = 1$ for convenience.

The pupil function P in the above formula determines the actual boundary of the integral, and when there is no obscuration:

$$P(r) = \begin{cases} 1, r \leqslant a \\ 0, r > a \end{cases} \tag{8.47}$$

where, a is the radius of the pupil. At this point, we can obtain the following formula by calculation:

$$U_3(r_3) = \frac{i\pi a^2}{\lambda f} \exp(-ikf)\exp\left(-\frac{i\pi r_3^2}{\lambda f}\right)\left[\frac{2J_1\left(\frac{2\pi r_3 a}{\lambda f}\right)}{\frac{2\pi r_3 a}{\lambda f}}\right]. \tag{8.48}$$

When the center of the objective is obscured, the pupil function becomes the form of a ring. Only the light at the annular part can pass through, and its function can be expressed as:

$$P(r) = \begin{cases} 1, & \varepsilon a \leqslant r \leqslant a \\ 0, r > a \text{ or } r < \varepsilon a \end{cases} \tag{8.49}$$

where, a is the outside diameter of the annular pupil; and ε is the ratio of the inside diameter to the outside diameter of the annular pupil.

The amplitude distribution on the back focal plane of the corresponding objective is as follows:

$$U_3(r_3) = \frac{i\pi a^2}{\lambda f} \exp(-ikf)\exp\left(-\frac{i\pi r^2}{\lambda f}\right)\left[\frac{2J_1\left(\frac{2\pi r_3 a}{\lambda f}\right)}{\frac{2\pi r_3 a}{\lambda f}} - \varepsilon^2\frac{2J_1\left(\varepsilon\frac{2\pi r_3 a}{\lambda f}\right)}{\varepsilon\frac{2\pi r_3 a}{\lambda f}}\right]. \tag{8.50}$$

The imaging spots under different obscuration ratios can be calculated by the above formula, and the modulation transfer function curves of the reflective objective under different obscuration ratios can be obtained by the Fourier transform of the spots. The modulation transfer function curves [4] under different obscuration ratios are shown in figure 8.10.

In figure 8.10, we can observe that the obscuration ratio of the system has significant effect on the modulation transfer function of the system. When there is no obscuration, the modulation transfer function curve of the system is the highest one, which indicates that the image quality is the best. With the increase in the obscuration ratio, the modulation transfer function curve goes down continuously. When the obscuration ratio is 5%, the modulation transfer function curve of the system with obscuration is slightly lower than that of the system without obscuration. When the obscuration ratio reaches 20%, the modulation transfer function curve of the system with obscuration is obviously lower than that of the system without obscuration; at the point where the normalized spatial frequency is 0.4, the

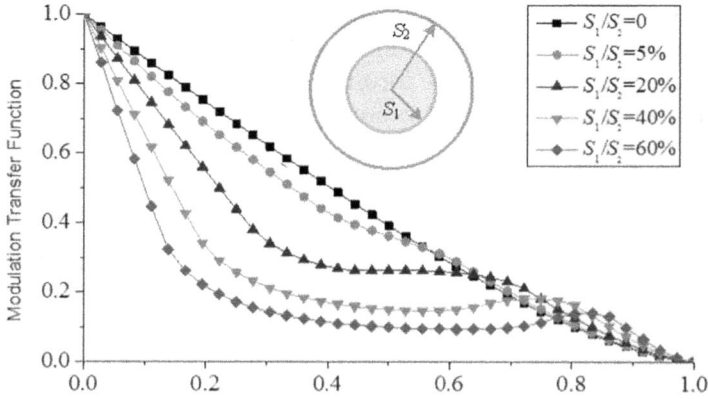

Figure 8.10. Modulation transfer function of different obscurations.

value of the modulation transfer function of the system without obscuration is about 0.5, while the value of the modulation transfer function of the system with obscuration is about 0.25, which is only half of that of the system without obscuration, indicating that the image quality declines obviously. When the obscuration ratio is increased to 60%, the value of the modulation transfer function of the system in the medium frequency band is only about 20% of that of the system without obscuration.

By the above analysis, we can draw the following conclusion: the obscuration ratio of the system has significant effect on the modulation transfer function of the system; the higher the obscuration ratio, the smaller the value of the modulation transfer function in low and medium frequency bands, and the worse the image quality. Therefore, to provide the imaging system having obscuration ratio with better image quality, the obscuration ratio shall be minimized.

8.5.2 Obscuration constraint model

The structure of the coaxial reflective objective determines that the obscuration is inevitable. Although obscuration cannot be eliminated, it can be minimized in the design, thus to reduce the effect of obscuration on system imaging. In the previous studies on reflective objective, more attention was paid to the characteristics of aberration, the obscuration ratio of the secondary mirror was not calculated and analyzed, and no uniform calculation model was established. The obscuration of aspherical reflective objective is analyzed in detail here, and a uniform obscuration calculation model of the secondary mirror of the aspherical reflective objective is established, thus to provide a theoretical basis for further optimizing the design of the aspherical reflective objective.

8.5.2.1 Obscuration constraint model

The structure of reflective objective is shown in figure 8.11, where the axial distance from object plane to the vertex of the primary mirror is ρ_0, the distance between the vertexes of the primary mirror and the secondary mirror is l_0, the axial distance from the vertex of the secondary mirror to the image plane is r_0, the radii of curvature of the primary mirror and the secondary mirror are R_1 and R_2, ρ is the optical path from an on-axis object point to the primary mirror, l is the optical path from the primary mirror to the secondary mirror, and r is the optical path from the secondary mirror to the image point. θ is the angle between the ray emitted from the on-axis object point and the optical axis, and u is the angle between the emergent ray and the optical axis.

In the reflective objective system, the entire optical path is symmetric about the optical axis, so all sections are circular. When ignoring the effect of the thickness of the secondary mirror, the obscuration ratio (OR) can be expressed as:

$$OR = \frac{S_0}{S_m} = \frac{\pi h_1^2}{\pi h_2^2}, \tag{8.51}$$

where, S_0 is the area of the secondary mirror; S_m is the total area of the light beam able to enter the optical system on the optical axis section where the secondary mirror is located; h_1 is the radius of the aperture of the secondary mirror; and h_2 is the radius of the light beam able to enter the optical system on the optical axis section where the secondary mirror is located.

It is known from the geometrical relationship that:

$$h_1 = r_0 \tan u. \tag{8.52}$$

$$h_2 = (\rho_0 - l_0)\tan \theta. \tag{8.53}$$

Therefore, the obscuration ratio can be expressed as a function of the basic structure parameters of the system:

$$OR = \frac{\pi h_1^2}{\pi h_2^2} = \left[\frac{r_0 \tan u}{(\rho_0 - l_0)\tan \theta}\right]^2. \tag{8.54}$$

The following approximation can be made when the angles q and u between the incident and emergent rays and the optical axis are small:

$$\frac{\tan u}{\tan \theta} \approx \frac{\sin u}{\sin \theta}. \tag{8.55}$$

It is known from the Abbe sine condition that:

$$\sin \theta = m \sin u, \tag{8.56}$$

where, m represents the magnification of the system;

Thus we can obtain:

$$OR = \left[\frac{r_0 \tan u}{(\rho_0 - l_0)\tan \theta}\right]^2 \approx \left(\frac{r_0}{\rho_0 - l_0}\right)^2 \cdot \frac{1}{m^2}. \tag{8.57}$$

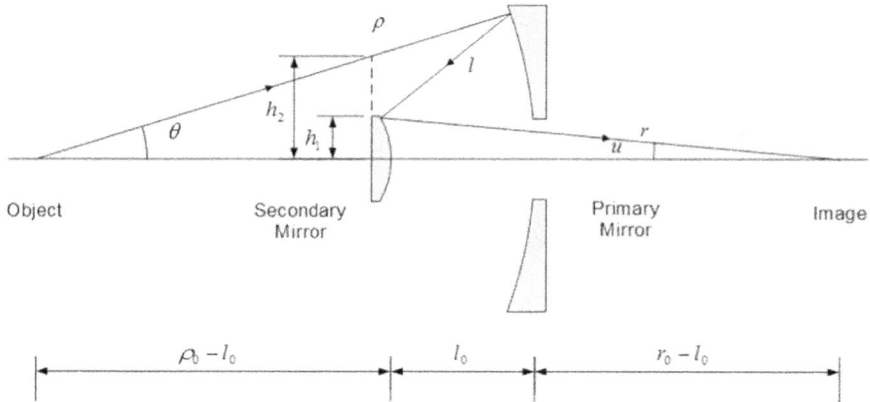

Figure 8.11. Calculation diagrams of obscuration.

Working distance is the axial distance from the object to the secondary mirror, and can be expressed as:

$$b = \rho_0 - l_0. \tag{8.58}$$

Therefore, the following function relationship among the obscuration ratio OR, the working distance b, the magnification m and the axial distance r_0 from the secondary mirror to the image plane is obtained:

$$OR = \left(\frac{r_0}{mb}\right)^2 = \left[\frac{r_0}{m(\rho_0 - l_0)}\right]^2. \tag{8.59}$$

Formula (8.59) is the obscuration ratio calculation model of aspherical reflective objective. We can observe that when the magnification m is constant, the obscuration ratio depends on the ratio of the axial distance from the secondary mirror to the image plane r_0 to the working distance b. When the working distance is increased or the axial distance r_0 from the secondary mirror to the image plane is decreased, the obscuration ratio is decreased. When the working distance is decreased or the axial distance r_0 from the secondary mirror to the image plane is increased, the obscuration ratio is increased. When the ratio of the axial distance from the secondary mirror to the image plane r_0 to the working distance b is constant, the higher the magnification is, the smaller the obscuration ratio is.

The key point of the secondary mirror obscuration model of the aspherical reflective objective is that it has established the constraint relation between the objective structural parameters and the obscuration ratio, thus satisfactory obscuration ratio can be obtained by adjusting the objective structural parameters. It shall be noted that during the derivation of the obscuration model, approximation is made in the condition of small angle, therefore when the angles q and u between the incident and emergent rays and the optical axis are large, the theoretical value of the obscuration ratio may be deviated from the actual calculation value.

8.5.2.2 *Total length calculation model*

The overall structure size of the optical system is an important index to be considered in the design of the optical system. When selecting the overall size, the specific situation and specific requirements of the application shall be considered. In figure 8.8, we can observe that the total length from the object plane to the image plane of the objective is as follows:

$$L = \rho_0 - l_0 + r_0. \tag{8.60}$$

By transforming the obscuration ratio calculation formula (8.36) and substituting formula (8.59) into formula (8.60), we can obtain:

$$r_0 = m(\rho_0 - l_0)\sqrt{OR}. \tag{8.61}$$

$$L = (m + 1)(\rho_0 - l_0)\sqrt{OR}. \tag{8.62}$$

In formula (8.62), we can observe that the total length of the mirror is in direct proportion to the working distance $(\rho_0 - l_0)$, the root of obscuration ratio $\sqrt{OR}$, and $(m + 1)$. To obtain a reflective objective with long working distance and small structure size, the magnification and obscuration ratio shall be decreased.

8.5.3 Design method based on obscuration constraint

In section 8.5.2, we have analyzed the obscuration ratio and established the constraint relation among the distance between the primary mirror and the secondary mirror of the reflective objective l_0, the axial distance from the object plane to the primary mirror ρ_0, the magnification m and the axial distance from the image plane to the secondary mirror r_0. While in section 8.4, we have obtained various decoupled parameter expressions of the objective. By combining both sections, we can easily obtain various structural parameters of a reflective objective with low obscuration ratio.

The parameters of the primary mirror of the aspherical reflective objective are sorted out as follows:

$$
\begin{cases}
OR = \left[\dfrac{r_0}{m(\rho_0 - l_0)}\right]^2, \\[2mm]
R_1 = -\dfrac{2ml_0\rho_0}{ml_0 + m\rho_0 - r_0}, \\[2mm]
k_1 = -\dfrac{(R_1 + \rho_0)^2}{\rho_0^2} - \dfrac{r_0 R_1^3(m + 1)}{m^2\rho_0^3(\rho_0 + r_0)}\left[\dfrac{\kappa^3(m + 1)^3(m - 1)^2}{4m(m\kappa - 1)^2(m - \kappa)^2}\right. \\[2mm]
\qquad - \dfrac{\kappa(m + 1)(m - 1)}{2(m\kappa - 1)(m - \kappa)} + \dfrac{\kappa(\kappa + 1)^2(m - 1)(2 - \alpha - \beta)}{4(m\kappa - 1)(m - \kappa)} \\[2mm]
\qquad \left. - \dfrac{(2 - \alpha - \beta)(\kappa + 1)(\alpha + m\beta - 1)}{2m} + \dfrac{(m + 1)(\alpha + m\beta)}{2m^2} - \dfrac{1}{m}\right].
\end{cases}
\tag{8.63}
$$

The parameters of the secondary mirror of the aspherical reflective objective are sorted out as follows:

$$\begin{cases} OR = \left[\dfrac{r_0}{m(\rho_0 - l_0)} \right]^2, \\[2ex] R_2 = \dfrac{2l_0 r_0}{r_0 + l_0 - m\rho_0}, \\[2ex] k_2 = -\dfrac{(R_2 - r_0)^2}{r_0^2} + \dfrac{\rho_0 R_2^3 (M+1)}{M^2 r_0^3 (r_0 + \rho_0)} \left[\dfrac{\kappa^3 (M+1)^3 (M-1)^2}{4M(M\kappa - 1)^2 (M-\kappa)^2} \right. \\[2ex] \quad - \dfrac{\kappa(M+1)(M-1)}{2(M\kappa - 1)(M-\kappa)} + \dfrac{\kappa(\kappa+1)^2(M-1)(2 - \alpha' - \beta')}{4(M\kappa - 1)(M-\kappa)} \\[2ex] \quad - \dfrac{(2 - \alpha' - \beta')(\kappa+1)(\alpha' + M\beta' - 1)}{2M} + \dfrac{(M+1)(\alpha' + M\beta')}{2M^2} - \left. \dfrac{1}{M} \right]. \end{cases} \qquad (8.64)$$

In the design of reflective microscope objective, first of all, we should determine the initial design data including objective working distance b, system obscuration ratio OR, system magnification m and the distance between the primary mirror and the secondary mirror l_0 according to the application situation and design requirements. Among which, the working distance b and the system magnification m are usually given in the design requirements; and the system obscuration ratio OR is the smaller the better, therefore a numerical value of not more than 10% can be set when setting parameters. In addition, the distance between the primary mirror and the secondary mirror l_0 shall make the objective size meet the requirements of the application situation.

Secondly, after determining the initial design data, we should calculate the axial distance from the object plane to the primary mirror ρ_0 and the axial distance from the secondary mirror to the image plane r_0 by the obscuration ratio calculation model of aspherical reflective objective deduced in section 8.5.2.

Thirdly, we should substitute the obtained reflective objective structural parameters ρ_0 and r_0 and the constants m and l_0 into the decoupled model (i.e. formulas (8.63) and (8.64)) deduced in section 8.5.3 to calculate the radii of curvature R_1 and R_2 and the conic coefficients k_1 and k_2 of the primary mirror and the secondary mirror. After the above steps, the design of the initial configuration parameters of the entire microscope objective is completed.

It should be noted that in the above method, the structure is formed on the basis of the quadric surface, and the effect of the high-order terms are ignored, so it is slightly different from the ideal surface. If you want to get the high-order terms, you can calculate it through the model in section 8.4.4, or you can optimize it directly through the optical simulation software Zemax. In addition, the thickness of the secondary mirror is ignored in the calculation. In fact, the working distance is the distance from the object to the back surface (plane) of secondary mirror, therefore the value of b shall be the working distance plus the thickness of the secondary mirror. However, this thickness is so small in relation to the working distance that it is ignored in theoretical derivation. In the application, the thickness of the secondary mirror can be given according to the actual situation, and optimization can be made by Zemax at the later stage of the design process.

8.6 Industrial application

The purpose of this chapter is to develop a microscopical instrument with extra-long working distance for assisting in the high precision stitching of multiple cameras. This instrument needs to meet the requirements of high magnification, multi-width spectrum, extra-long working distance, and compact and simple structure. Its specific parameters are: the overall magnification of 16, the working distance of 525 mm, and the working spectrum of 400–900 nm. In order to reduce the size of the instrument as much as possible, if a purely reflective objective is used to meet the above requirements, as the object distance is very large, the image distance will be increased exponentially with the increase of the magnification, causing the oversize of the overall structure. Therefore in design, the method of two-stage amplification shown in figure 8.12 is used to meet the requirements of magnification and structure size. The image plane side of the reflective objective is butted with the second-stage microscopic module for second-stage amplification, and the overall magnification of the system is the product of the magnification of the reflective objective at the front end and the magnification of the second-stage microscopic module. Since there are no requirements for the working distance on the image plane side of the reflective objective, the second-stage microscopic module can be selected from commercial apochromatic objective products. The numerical aperture of the objective of the second-stage microscopic module is required to be greater than the image numerical aperture of the reflective objective to ensure that the image quality will not be decreased. This section gives the specific requirements of the reflective objective, as follows: working distance of 525 mm, the magnification of the objective of 6.5, the objective numerical aperture of 0.13, and the obscuration ratio of the system set to 4%. The second-stage microscopic module is selected from mature commercial products. For example, we use Zeiss EC Plan Neofluar (NA = 0.075, magnification is 2.5) as the objective, and the selection criteria are that the apochromatism capacity is high, the spectrum covers 400–900 nm (the transmittance is higher than 80%) and the objective numerical aperture is greater than the image numerical aperture of the first-stage aspherical objective; use Mitutoyo VMU-V as the tube mirror, and the reason for selecting this tube mirror is that the module has an achromatism of 400–1800 nm and is compact and easy to install; use a Sony EI-30CE near infrared camera as the CCD because this camera has a wide working band and high sensitivity, and its working wavelength spectrum covers 400–900 nm; use an Edmund Optics F39-481 beamsplitter with a thickness of 2 μm as the film beamsplitter because this film is very thin and thus can eliminate the ghosting, avoid additional aberration of the optical system caused by the beamsplitter and be applicable to visible light and near infrared wave band; use the customized LED light source produced by the OPT company as the light source because this LED light source has the output wavelength of 480 nm, 550 nm, 660 nm and 850 nm and

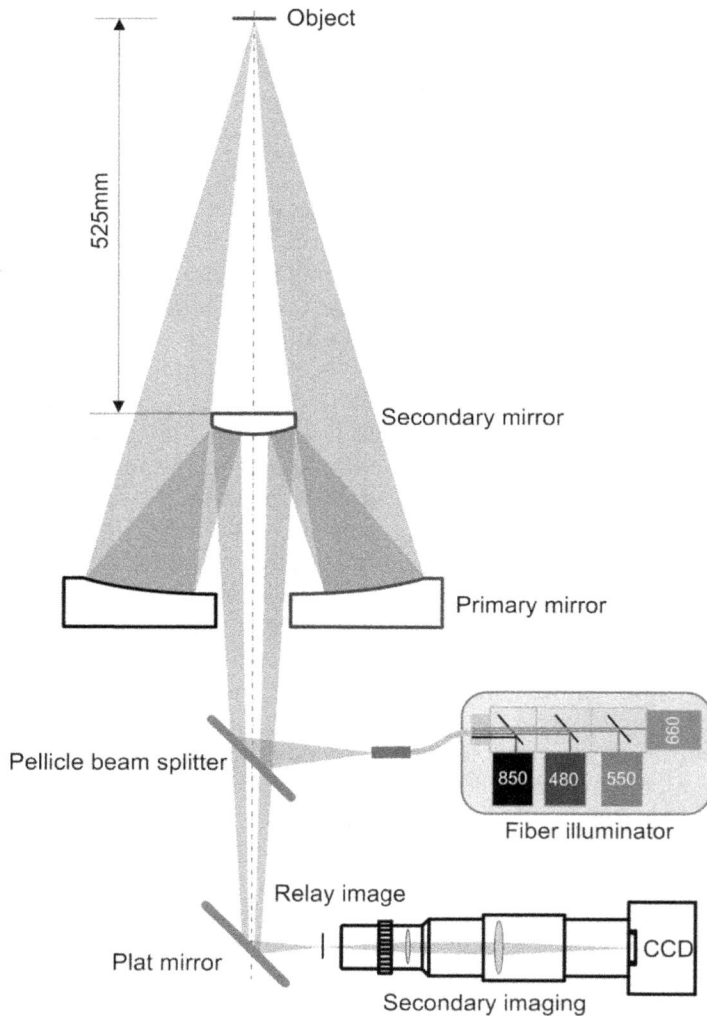

Figure 8.12. Schematic diagram of microscope with super long working distance [5].

is characterized by small calorific value, optional wavelength, energy concentration, high luminous efficiency, flexible application, high output intensity and adjustable output light intensity; and use a Mortex MSG4-2200-HR optical fiber for light transmission because its transmittance at the band of 400-900 nm is about 50%, which can meet the working requirements of microscope. We have adopted a staged model, and the second-stage microscopic module is selected from mature commercial products, so in addition to the design idea, the most important thing in the development of the instrument is the design of the first-stage aspherical reflective objective.

8.6.1 Structural parameter calculation

First of all, calculate the three distances of the reflective aspherical objective, i.e. the axial distance from the object plane to the primary mirror, ρ_0, the distance between the primary mirror and the secondary mirror, l_0, and the axial distance from the secondary mirror to the image plane, r_0. l_0 is the distance between the two mirrors. Considering the overall structure size, we take l_0 as 190 mm. The thickness of the secondary mirror is 25 mm, i.e. $b = 550$ mm. According to the equation $\rho_0 = l_0 + b$, we can obtain $\rho_0 = 740$ mm. It can be calculated from formula (8.51) that the axial distance from the secondary mirror to the image plane is as follows:

$$r_0 = mb\sqrt{OR} = 715 \text{ mm}. \tag{8.65}$$

After the three distances ρ_0, l_0, r_0 and the system magnification m are obtained, there are only two surface shape parameters of the aspherical surface unknown. Substitute ρ_0, l_0, r_0 and m into the model of aspherical surface in rectangular coordinate system obtained by formulas (8.63) and (8.64) to calculate the radii of curvature and conic coefficients of the primary mirror and the secondary mirror respectively, and the results are shown in table 8.3. During calculation, only the conical coefficient among the aspherical coefficients of the primary mirror and the secondary mirror is calculated, and the high-order aspherical coefficient is not calculated.

The initial configuration parameters of the reflective objective obtained in the final design are shown in table 8.4. Among which, the numerical aperture of the objective is 0.13, the magnification is 6.5, and the wavelength is set as 480 nm, 550 nm, 660 nm and 850 nm.

Evaluate the image quality of the above initial configuration to see if it meets the imaging requirements. It shall be noted that during the derivation in this section and this chapter, the final parameter formulas are obtained by ignoring the high-order

Table 8.3. Parameters of a primary mirror and secondary mirror.

Parameters	Values
Radius of curvature of primary mirror, R_1	−342.927 mm
Conic coefficient of primary mirror, k_1	−0.273 86
Radius of curvature of secondary mirror, R_2	−69.578 mm
Conic coefficient of secondary mirror, k_2	−0.963 52

Table 8.4. Parameters of the designed reflective objective.

Type	Radius (mm)	Thickness (mm)	Conic coefficient	Half aperture (mm)
Object plane	Infinity	740	0	0.3
Primary mirror	−342.927	−190	−0.27386	95.3
Secondary mirror	−69.578	715	−0.96352	14.4
Image plane	Infinity	–	–	1.95

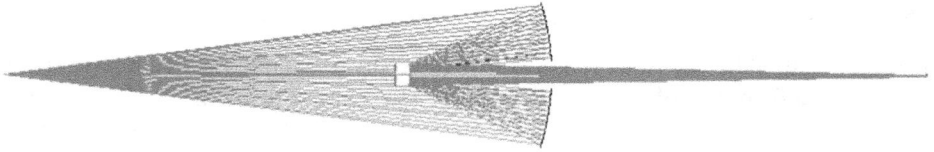

Figure 8.13. Structure of designed reflective objective.

items, therefore the field of view at the edge and the truncation error of high-order aspherical surface have inevitably produced aberration. In order to reduce the system aberration, high-order aspherical coefficient can be introduced because the high-order coefficient will retain more series terms, thus to reduce the error of using general aspherical function approximation instead of ideal aplanatic aspherical surface. In this case, it shall be noted that, with the increase of retained terms, the complexity of theoretical calculation will increase rapidly, which will be a key factor in the design and make it more difficult for designers. Therefore, the two aspects shall be balanced in the actual design.

8.6.2 Obscuration verification

Figure 8.13 shows the structure of reflective objective obtained by Zemax simulation. The object is on the same side as the secondary mirror, and the image is on the same side as the primary mirror. The imaging beam is obscured by the secondary mirror, passes through the primary mirror and the secondary mirror successively, and finally forms an image in the image plane. When the size of the selected CCD image plane is 3.6 mm × 4.8 mm and the magnifications of the reflective objective at the front end and the microscopic module at the back end are respectively 6.5 and 2.5, the actual corresponding object space field of view of the microscope is 0.22 mm × 0.3 mm; when the height of object space field of view is set as 0, 0.1 mm, 0.2 mm and 0.3 mm, the simulated corresponding maximum field of view is 0.6 mm, which is far greater than the observable field of view of the microscope.

By simulation, we can obtain that the aperture of the secondary mirror is 14.4 mm and the aperture of the beam in the plane where the secondary mirror is located is 72 mm. When effect of the mechanical support structure is not considered, the obscuration ratio of the system is as follows:

$$OR = \left(\frac{h_1}{h_2}\right)^2 = 4\%. \tag{8.66}$$

The simulated obscuration ratio is consistent with the designed obscuration ratio, the designed obscuration requirements are met, and the correctness of the reflective objective design method based on obscuration constraint is verified.

8.6.3 Analysis of initial configuration imaging characteristics

The parameters of reflective objective in table 8.4 are obtained by the reflective objective design method based on obscuration constraint. It is necessary to evaluate the image quality before processing in order to determine whether the initial configuration meets the imaging requirements. There are various methods for image quality evaluation, and the common methods are wave aberration OPD, spot diagrams Spt and optical transfer function. Wave aberration is applicable to the evaluation in the phase of optical system design; Spt is used for detecting the actual image quality of the product; and optical transfer function is an evaluation criterion applicable to both design and manufacturing phases. In this section, the imaging characteristics of the reflective objective designed in section 8.6.1 is analyzed in accordance with the evaluation criteria of wave aberration and optical transfer function.

The spherical waves emitted from the object point are still focused ideal spherical waves after the imaging of the ideal system. When the optical system has aberration, the focused spherical waves will deviate from the ideal spherical waves, and the deviation of the actual wave surface relative to the ideal spherical waves is the wave aberration. The magnitude of wave aberration reflects the image quality of the optical system. The smaller the wave aberration, the closer the image is to the ideal image and the better the image quality. According to Rayleigh criterion, when the maximum deviation between the actual wave surface and the reference spherical surface, i.e. the wave aberration, is less than 1/4 wavelength, the actual wave surface can be considered to be nondefective.

Figure 8.14 shows the OPD of reflective objective with initial configuration. It can be used to show the wavefront deformation of the light emitted by an object during transmission in the objective system. In the figure, the four subgraphs are respectively corresponding to four fields of view, and the wave aberrations in the meridian plane and the sagittal plane are evaluated for each field of view. The

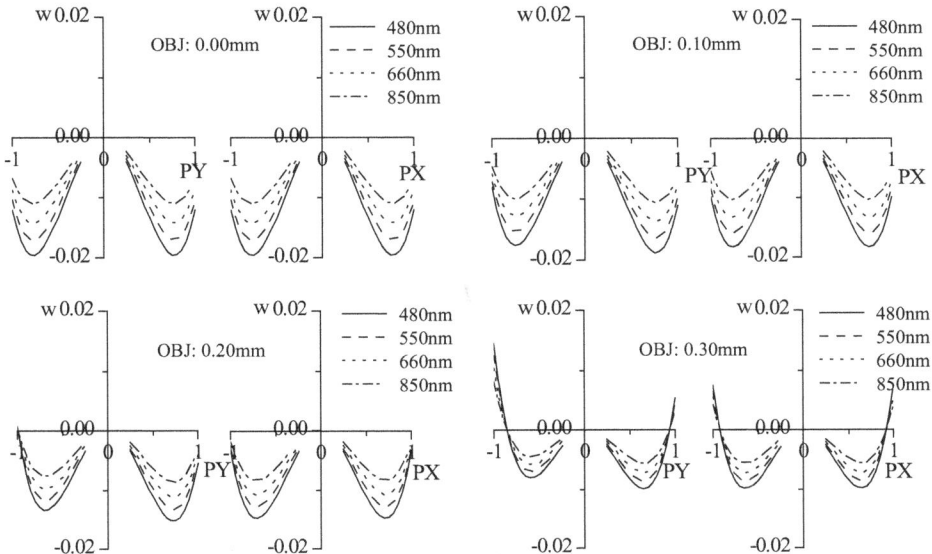

Figure 8.14. OPD of objective with initial configuration.

horizontal axes PY and PX are the normalized coordinates of the entrance pupil of the optical system, indicating the meridional direction and the sagittal direction respectively; the vertical coordinate W is the wave aberration at the position of the exit pupil, and the unit is the wavelength.

In figure 8.14, we can observe that, as the wavelength increases, the wave aberration decreases. The corresponding wave aberration of 480 nm is the largest, and the corresponding wave aberration of 850 nm is the smallest. This is because the wave aberration is measured in wavelength. If the optical path difference is the same, when short wavelength is converted into a quantitative value in wavelength, the corresponding quantitative value is high. The wave aberration of the system in each of the four fields of view is less than 0.02 wavelength, which is far less than 1/4 wavelength according to Rayleigh criterion, therefore the system has good image quality from the perspective of wave aberration.

Optical transfer function is a uniform image quality evaluation criterion, which is applicable to both design and manufacturing phases. The optical transfer function is to carry out Fourier transform to the light field distribution function of the object and study the information carrying capacity of the optical system for various spatial frequencies. The low frequency part reflects the optical system's capacity to carry the contour information of the object, the medium frequency part reflects the optical system's capacity to carry the hierarchical information of the object, and the high frequency part reflects the optical system's capacity to carry the detailed information of the object. The optical transfer function includes two parts, which are modulation transfer function and phase transfer function. For imaging, the amplitude is decisive, therefore only modulation transfer function (MTF) is calculated in general. In the actual operation, the actual modulation transfer function of the optical system is usually contrasted with the diffraction limit modulation transfer function of the system, and the deviation of the two functions determines the actual image quality of the system. The closer the actual modulation transfer function is to the diffraction limit modulation transfer function, the better the object information carrying capacity is, and the better the image quality.

Figure 8.15 shows the modulation transfer function curve of the designed reflective objective and the modulation transfer function curve of the ideal optical system without obscuration. The lateral coordinate of the curve is the spatial frequency in (lp/mm), and the vertical coordinate is the contrast ratio of the output image to the input image. It is the evaluation of the image space here, so it is corresponding to the modulation transfer function curve of the image space. It can be approximately considered that the only difference between the modulation transfer function curve of the object space and the modulation transfer function curve of the image space is the magnification of the lateral coordinate.

In figure 8.15, we can observe that the two overlapping curves at the top are the modulation transfer function curves of the ideal optical system without obscuration, and the curves at the lower part are the modulation transfer function curves of the designed aspherical reflective objective. The corresponding modulation transfer function curves of each field of view of the designed aspherical reflective objective completely coincides with the modulation transfer function curve of diffraction limit

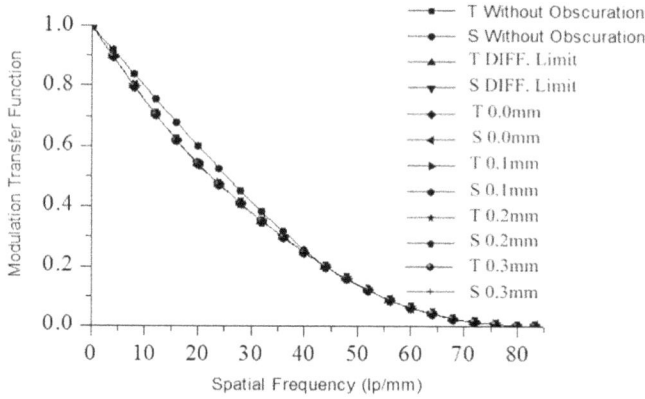

Figure 8.15. The contrast of modulation transfer function of objective in imaging space.

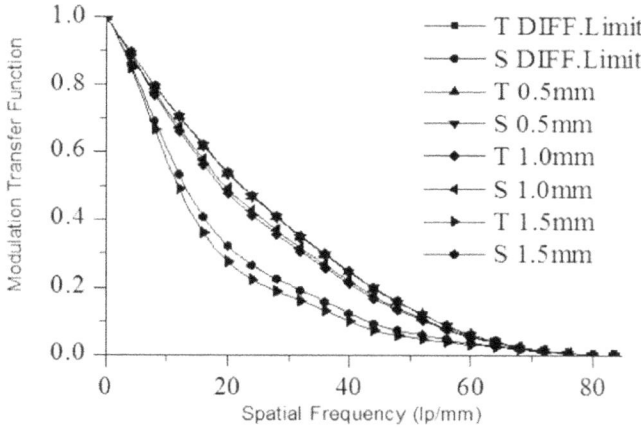

Figure 8.16. Modulation transfer function of objective with large field.

of the objective itself, indicating that the imaging result of the reflective objective is close to the ideal imaging result in the condition that the semifield is 0.3 mm; while the modulation transfer function curve of the designed objective has a small deviation from the modulation transfer function curve of the ideal optical system without obscuration, indicating that the obscuration ratio of the designed objective is low and the design goal of reducing obscuration ratio is achieved.

Figure 8.16 shows the modulation transfer function curves corresponding to each field of view of the designed reflective objective in the condition that the field of view of the reflective objective is further increased. The heights of the fields of view of the reflective objective are respectively set as 0, 0.5 mm, 1 mm and 1.5 mm, and the other conditions are unchanged. In figure 8.16, we can observe that the modulation transfer function curve corresponding to the 1 mm field of view still basically coincides with the modulation transfer function curve of diffraction limit, indicating that the image quality of the reflective objective is good in this field of view. When the field of view is further increased, the deviation of the corresponding modulation transfer function curve from the modulation transfer function curve of diffraction limit is increased, indicating that the image quality is decreased in the field of view at the edge.

According to the above image quality analysis, the microscope objective with long working distance designed by the reflective objective design method based on obscuration constraint meets the requirement of long working distance and has low obscuration ratio and small aberrations, and its image quality is close to the ideal image quality.

8.6.4 Whole equipment and specific application

The aspherical surfaces of the primary mirror and the secondary mirror of the reflective objective are processed by traditional optical processing method and made of fused quartz. The primary mirror has a thickness to radius ratio of about 1:8, an aperture of 210 mm and a hollow aperture of 36 mm; the secondary mirror has an aperture of 29 mm. The central area of the secondary mirror is not involved in imaging and is blackened in order to reduce the stray light, and the back surface is polished for the convenience of installation and adjustment. The processing surface chooses protected aluminum film which has a good reflectivity in 0.4–20 μm and meets the requirement of design band (400–900 nm). Use an interferometer to test the primary mirror and the secondary mirror, and their surface shape PV values are 85 nm and 51 nm respectively, which are less than 1/4 wavelength (632.8 nm) and meet the processing requirements.

First, image the resolution target to evaluate the imaging characteristics of the produced instrument. In the experiment, place the 1951USAF resolution target near the object plane of the instrument, and accurately adjust the position of the resolution target relative to the object plane of the microscopical instrument by a Z-direction one-dimensional adjustment mechanism to make the resolution target clearly image on the CCD. During the experiment, keep the position of the object plane unchanged, connect LED light sources with four wavelengths ($\lambda = 480$ nm, 550 nm, 660 nm and 850 nm) successively by optical fiber, and observe the imaging results of the microscopical instrument for the resolution target with different wavelength light sources. Figure 8.17 shows the imaging results of the produced microscopical instrument for the 1951USAF resolution target with different wavelength light sources, and the corresponding lighting wavelengths of subgraphs (a)–(d) are 480 nm, 550 nm, 660 nm and 850 nm.

The position of the minimum number of line pairs shown in figure 8.17 is the seventh element of Group 7 (228.1 lp/mm) of 1951USAF resolution target. The microscope can clearly image the resolution target with different wavelength light sources at the same object plane position, indicating that the system has small axial chromatic aberration and small other aberrations. In addition, as the wavelength increases, the contrast decreases. This is because the resolution of the system is in

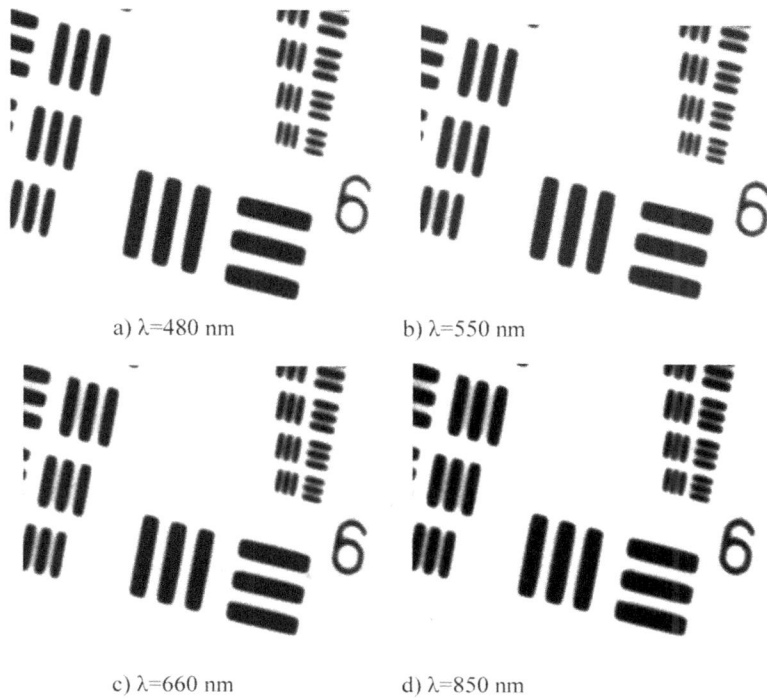

a) λ=480 nm   b) λ=550 nm

c) λ=660 nm   d) λ=850 nm

Figure 8.17. Imaging results of 1951USAF resolution target with different wavelength light sources.

inverse proportion to the wavelength, and the larger the wavelength, the lower the resolution (Airy disk diameter $D = 1.22\lambda/\text{NA}$).

According to the actual measurement, the distance from the front face of the mechanical cover of the microscopical instrument to the resolution target is 525 mm, i.e. the working distance of the microscopical instrument achieves 525 mm; while the numerical aperture NA depends on the reflective objective at the front end, i.e. NA = 0.13. Place the resolution target on the XY two-dimensional objective table, and the field of view of the microscopical instrument can be demarcated. When the demarcated CCD is Sony EI-30CE, the object space field of view of the produced microscopical instrument is about 0.30 mm × 0.22 mm, the size of the CCD image plane is 4.8 mm × 3.6 mm, and a magnification of about 16 can be obtained, which is consistent with the design value.

Secondly, observe a linear array camera under the produced instrument to test the actual imaging performance of the microscope system for the sample with the complex structure. The sample is Thorlabs linear array camera LC100 with a single pixel size of 14 μm × 56 μm and a working wavelength of 350–1100 nm.

In the experiment, place the linear array camera near the object plane of the microscopical instrument, and accurately adjust the position of the linear array camera relative to the microscopical instrument by a Z-direction one-dimensional adjustment mechanism to make the linear array camera clearly image on the CCD.

At this moment, keep the position of the object plane unchanged, connect LED light sources with four wavelengths ($\lambda = 480$ nm, 550 nm, 660 nm and 850 nm) successively by optical fiber, and observe the imaging results of the microscopical instrument for the linear array camera with different wavelength light sources. Figure 8.18 shows the imaging results for the linear array camera with different wavelength light sources at the same axial position.

In figure 8.18, we can observe that the produced microscopical instrument can clearly image the line CCD with different wavelength light sources at the same axial position, indicating that the image quality of the microscopical instrument for the sample with complex structure is good, and verifying that the chromatic aberration of the microscopical instrument is good. As the wavelength increases, the contrast of the sample decreases gradually. This is consistent with the imaging phenomenon of the resolution target and is because the resolution of the system will decrease along with the increase of the wavelength.

The designed microscopical instrument with extra-long working distance can clearly image the resolution target and the complex-structure sample with different wavelength light sources at the same axial position, indicating that the system has good image quality and apochromatism within the lighting wavelength and achieves the design goal.

The overall framework of the instrument is made of marble to reduce the effect of vibration and temperature. Some of the mechanical clamping parts and structures in the instrument have been processed accordingly, which will be not repeated herein. The whole instrument is shown in figure 8.19.

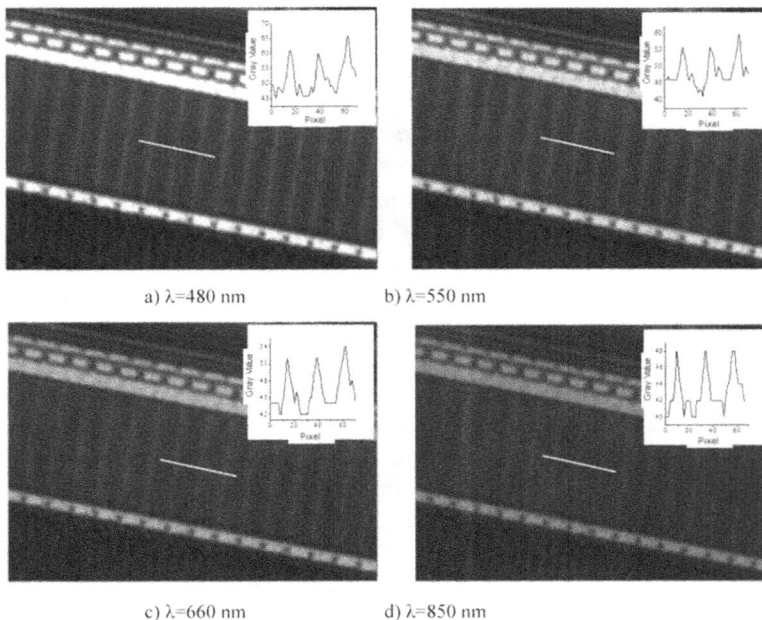

a) λ=480 nm   b) λ=550 nm

c) λ=660 nm   d) λ=850 nm

Figure 8.18. Imaging results of line CCD with different wavelength light sources.

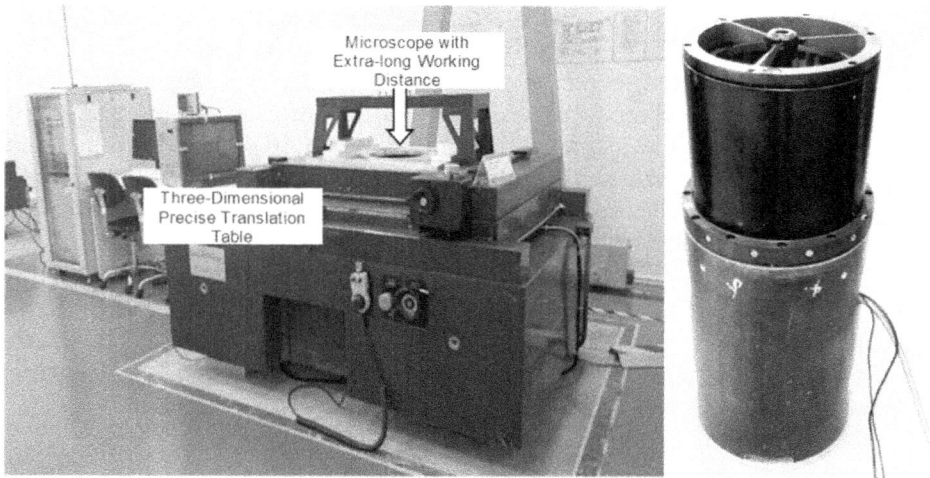

Figure 8.19. Designed instrument.

8.7 Summary

In this chapter, first of all, two simple and effective aspherical mirror design models are proposed to answer the question of 'how do we design an aspherical mirror'. Second, we answered the question of 'how do we build an aspherical reflection system' by actually designing an aspherical reflective imaging system. The main contributions of this chapter are summarized as follows:

1. An analytical design model of aspherical reflective objective in rectangular coordinate system is established. This model can be used to calculate the structure parameters of the aspherical reflective objective simply and effectively, which solves the problem that the curvature radius of aspheric surface's vertex and the aspherical coefficient cannot be decoupled. Simulation results show that when the numerical aperture of the aspherical reflective objective is less than 0.5, the imaging quality of this reflective objective whose structure parameters are calculated by this model is close to diffraction limit.

2. A design model of aspherical reflective objective based on obscuration constraint is proposed. This model establishes the constraint relation model between the structural parameters of the aspherical reflective objective, solves the problem that the obscuration caused by the secondary mirror is large in the design of the reflective objective. Simulation results show that the design model can reduce the obscuration ratio to less than 5% without affecting the imaging effect.

3. An aspherical reflective imaging system (extra-long working distance microscopy) is designed and developed, and the experiment is carried out on this instrument. According to the proposed design model, we have successfully developed an aspherical reflective objective with the working distance of

525 mm, the numerical aperture of 0.13, the magnification of 6.5 and the obscuration ratio of 4%, and combined the secondary microscopic module to implement a microscopic instrument with long working distance, wide spectrum, large magnification. The working distance of the microscope is up to 525 mm, the magnification rate is up to 16 times, the range of the imaging spectral bands is 400–900 nm and the 7th element of group 7 of the 1951 USAF identification rate board (the resolution level is 228.1 lp/mm) can be clearly visualized.

References

[1] Schwarzschild K 1905 Göttingen. Untersuchungen Zur Geometrischen Optik. III *Astronomische Mittheilungen der Königlichen Sternwarte zu Göttingen* (Göttingen: Universität Göttingen) pp 11–5

[2] Head A K 1957 The two-mirror aplanat *Proc. Phys. Soc. Sect.* B **70** 945–51

[3] Artyukov I A 2012 Schwarzschild objective and similar two-mirror systems *Proc. SPIE* 2012 86780A:1–5

[4] Smith W J 2005 *Modern Lens Design* (New York: McGraw-Hill) pp 567–80

[5] Tan J, Wang C, Wang Y and Liu J *et al* 2014 Long working distance microscope with a low obscuration aspherical Schwarzschild objective *Opt. Lett.* **39** 6699–702

[6] Korsch D 1991 *Reflective Optics* (San Diego, CA: Academic)

Chapter 9

Elliptical mirror applied in TIRF microscopy

Qiang Li, Mengzhou Li, Chenguang Liu, Jian Liu and Jiubin Tan

9.1 Introduction

As one of the earlier applications of the theory, we propose and demonstrate an elliptical mirror-based TIRF (e-TIRF) microscopy with shadowless illumination and adjustable penetration depth [1]. The elliptical mirror is used to produce a hollow cone illumination with all azimuthal directions and a large range of incident angle, so as to attenuate the potential shadow effects when utilizing a single direction illumination, such as asymmetries and low contrast. This application demonstrates that the e-TIRF method makes an illuminated sample almost shadowless with symmetric intensity distribution. Meanwhile, penetration depth is theoretically adjustable from 58 nm to 250 nm, because incident angles are changeable by adjusting the size of the aperture or the position of an opaque mask. This method extends the minimum penetration depth, which is useful for high axial resolution.

9.2 Background

As a sort of aspherical mirror, an elliptical mirror has been widely used in the manufacture of optical microscopes, telescopes and so on. We have discussed about the basic theory for a high NA system and simulation in previous chapters. Here, we experimentally employ an elliptical mirror in one of the most popular microscopy, total-internal-reflection fluorescence (TIRF) microscopy. TIRF microscopy is a strong tool for imaging adherent cells [2] with excellent optical sectioning capability [3] and high signal-to-noise ratios (SNRs) [4], whose typical applications are about single molecule imaging [5], cellular endocytosis processing [6], quantitative measurements of surface protein density [7] etc.

Prism-based TIRF microscopy is an early developed TIRF approach [8]. Generally, a prism is involved to make illumination beam coupled into the interface for realizing total internal reflection (TIR). The maximum incident angle will thus be determined by the prism geometrical structure. Objective-based TIRF microscopy is another popular technique. In comparison with prism-based TIRF microscopy, it is more compacted and simple in operation [9]. In this method, a high NA objective is used to offer high angle illumination, but this illumination is still a single direction. However, as a result of single direction illumination, shadows along the beam path are unavoidable as the illumination direction is not multiple [10]. The issue of shadows or other artifacts is to attenuate image contrast [11], and resolving the ability gets lower as well. In order to deal with shadow problems, A L Mattheyses *et al* employed a rotating optical wedge to produce a hollow cone beam in time domain, with a focused spot at the back focal plane of objective to realize effectively uniform illumination [12]. The interference fringes and anisotropic blur are averaged and suppressed during fast scanning process, but the incident angle is fixed. W Zong *et al* proposed a siva-TIRF method in 2014. Siva-TIRF method is another alternative to overcome shadows by six positions scanning [13]. Moreover, M Lei and A Zumbusch's utilized W-shaped axicon mirror to weaken shadows, but illumination angle is merely adjustable in a small range [14].

Recently, penetration depth control attracts more attention because TIRF microscopy with photobleaching has succeeded in reconstructing a three-dimensional (3D) image with axial superresolution and molecule localization [15]. The photobleaching process needs different evanescent field penetration depths altered by tuning the incident angle of the excitation light. Nevertheless, the penetration depth of objective-based TIRF microscopy is almost fixed and restricted by objective aperture angle [9]. In order to control penetration depth, K Stock *et al* investigated on the VA-TIRF method and change incident angle by using a hemi-cylindrical prism in a compact device [16]. The penetration depth of an evanescent field is permissively tunable by changing the incident angle, whereas the device assembling with a number of fine mirrors is very difficult and expensive. In addition, galvo mirrors are also used to change incident angle [16]. The maximum incident angle is around 65.98°, still under the maximum aperture angle.

In this chapter, we present an e-TIRF microscopy with shadowless illumination and adjustable penetration depth, which is similar to a shadowless lamp. The elliptical mirror offers a hollow cone illumination with all azimuthal illumination directions and a large range of incident angle of nearly 30°. Illumination angle can be controlled by adjusting aperture size and the position of an opaque mask.

9.3 Basic theory

The experimental setup is demonstrated in figure 9.1 A collimated, linearly polarized 532 nm laser beam (MW-SGX, 532 nm, CLASS IIIb) propagates through two sets of 4f-systems, and is focused to a hollow cone beam by the condenser lens. The first 4f-system, consisting of the lens L1 ($f = 40$ mm), lens L2 ($f = 200$ mm), mask 1 ($D = 2$ mm) and an aperture, reshapes the collimated laser beam to a hollow expanded annular beam with inner and outer radius adjustable. Mask 1 is placed in front of L2 and obstructs the central part of the beam. Adjusting the position of mask 1 along the optical axis can control the inner radius of the ring beam, while changing the window radius of the aperture behind L2 modifies the outer radius. Mask 2 is used for limiting the minimum inner radius. The second 4f-system consists of lens

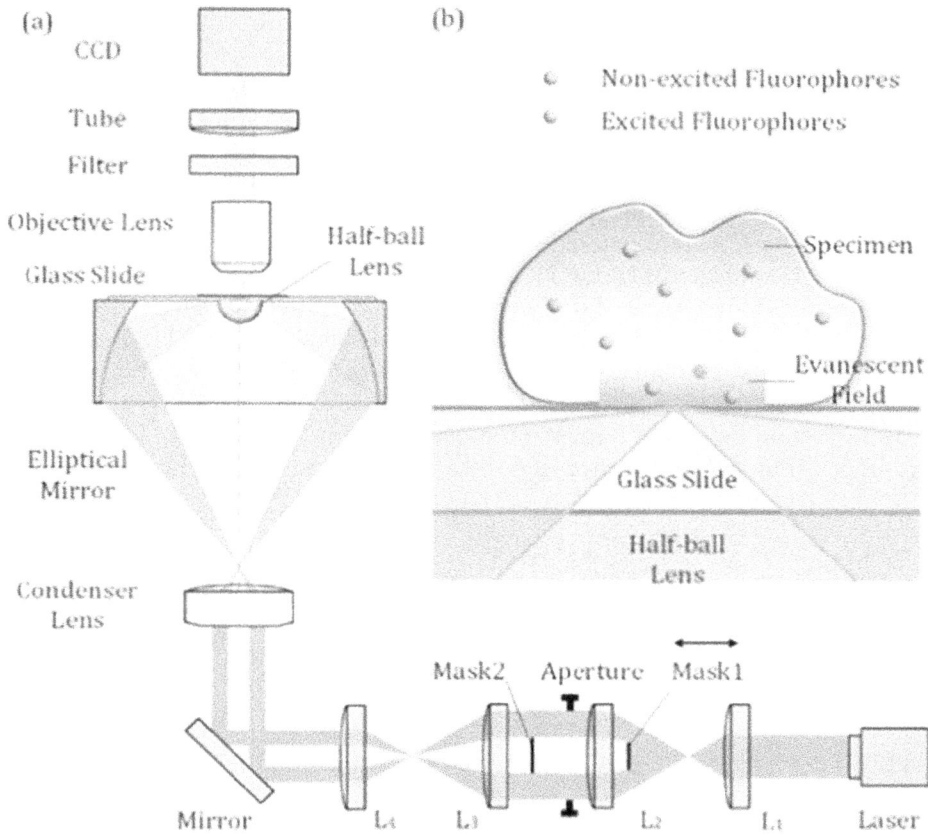

Figure 9.1. The schematic diagram of e-TIRF microscopy [1]. (a) The scheme of e-TIRF system construction. (b) The process of fluorescence excitation in e-TIRF microscopy.

L3 ($f = 200$ mm) and L4 ($f = 100$ mm), and shrinks the annular beam to a proper size compatible to the entrance pupil of condenser lens. The lower focus of elliptical mirror coincides with the focus of condenser lens. Then the elliptical mirror collects the hollow cone beam from condenser lens and refocuses it to the center of the half-ball lens ($D = 5$ mm, Edmund Optics). As the key part of this system, the upper end face of the home-made elliptical mirror serves as the reference plane of its upper focus. Therefore, the elliptical mirror can focus light on the lower surface of specimens when the specimens are placed on the upper end face of it. The half-ball lens, directing light to ensure total internal reflection, is stuck on the lower surface of glass slide which is placed on the focus plane. The half-ball lens is glued to the glass slide by the microscope immersion oil (Low Autofluorescence Immersion Oil, $n = 1.518$, Olympus Type F) matching the refractivity of the glass slide and the half-ball lens. A long-pass filter (562 nm, 25.0 mm Diameter, Dichroic Filter, Edmund) blocks the excitation laser beam. The incident angle range is controllable as shown in figure 9.1(b). The process of fluorescence excitation is shown in figure 9.1(b). The specimen contacted to the surface of glass slide is excited by a hollow cone beam.

Shadows can be attenuated by shadowless illumination, consisting of all azimuthal illumination directions mainly and a large incident angle range as well as efficient illumination. For conventional ways, prism-based TIRF setup produces a single illumination direction in figure 9.2(a). The range of incident angle in prism-based TIRF method (20 mm, N-BK7, Right-Angle Prism, Thorlabs) is from 61.18° to 72.76°, as depicted schematically in figure 9.2(b). The critical angle of TIR is 61.18° when the refractive index in both sides of the surface are 1.33 and 1.518. Meanwhile, objective-based TIRF equipment also offers a single illumination direction in figure 9.2(c). The incident angle in objective-based TIRF method is conditional on NA, even when the objective has a high NA (maximum angle 78.98° for 1.49 NA with refractive index 1.518 of immersion oil, and 67.97° for 1.65 NA with refractive index 1.78 of immersion oil) as shown in figure 9.2(d). In the e-TIRF microscope, samples are excited by the hollow cone beam with all azimuthal

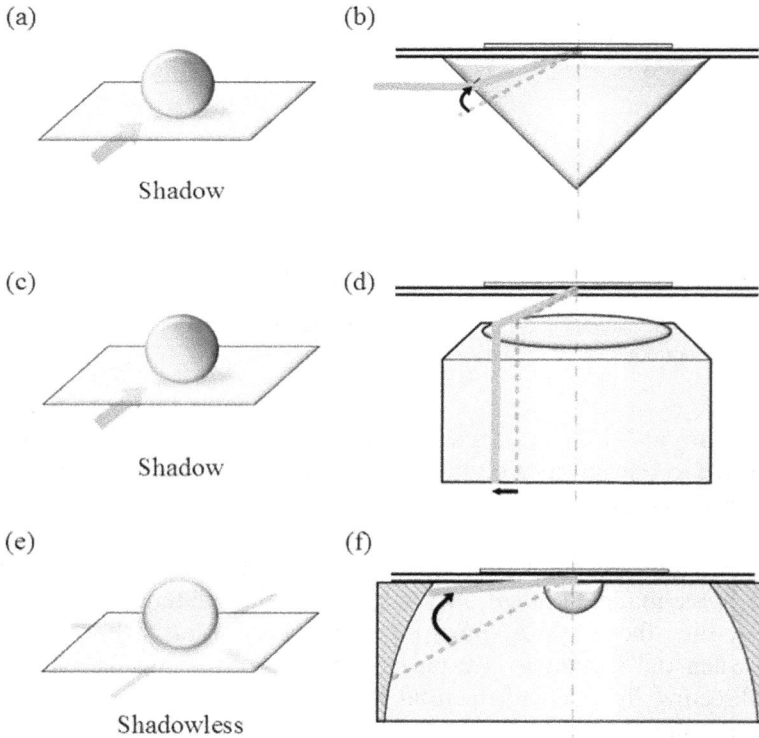

Figure 9.2. Comparison of three TIRF methods [1]. (a) A single illumination direction produces shadows in prism-based TIRF method. (b) The range of incident angle in prism-based TIRF method. (c) A single illumination direction produces shadows in objective-based TIRF method. (d) The range of incident angle in objective-based TIRF method. (e) Multiple illumination directions weaken shadow in e-TIRF method. (f) The range of incident angle in e-TIRF method.

directions in figure 9.2(e). Incident angle changes from critical angle to almost 90°, whose range is more than twice as much as other traditional methods in figure 9.2(f).

The penetration depth (d) of evanescent field in TIRF microscopy is determined by the incident angle (θ) as shown in the following formula:

$$d = \frac{\lambda}{4\pi(n_1^2 \sin^2 \theta - n_2^2)^{1/2}},$$

where n_1 and n_2 are the refractive index of the glass slide and the specimens respectively and the value of n_1 and n_2 are 1.518 and 1.33 normally. How θ influences on d in three TIRF methods is shown in figure 9.3. The bottom area marks out the effective excitation region and the blue curve represents the relationship of θ–d in TIRF microscopy. In addition, the red dashed line signs the critical angle of total internal reflection between n_1 and n_2. By comparing diagram, θ is in a range of 61.18° to 72.76° in prism-based TIRF method and ranges from 61.18° to 78.98° in objective-based TIRF method, while it changes from 61.18° to nearly 90° in e-TIRF method. Correspondingly, the maximum d is 250 nm, while the minimum d is theoretically 73 nm, 63 nm and 58 nm, respectively. For other prisms, the incident angle is still restricted by geometric structure in prism-based TIR. For high NA objective (1.65 NA), d is lower than e-TIRF above because of high immersion oil in objective-based TIRF. We can get similar results if the half-ball lens and glass slide have high refractive index in e-TIRF.

9.4 Experiments

To demonstrate the elliptical mirror-based TIR method's character, we image general microspheres (Lumisphere, 5 μm, Polystyrene, BaseLine) under widefield illumination with a fiber optic illuminator (115V, MI-150, Edmund, America) as shown in figure 9.4(a) and under three different illumination modes with a laser as shown in figures 9.4(b)–(d) respectively. The first mode utilizes propagating light to illuminate straightway in figure 9.4(b). In the second mode prism-based TIR

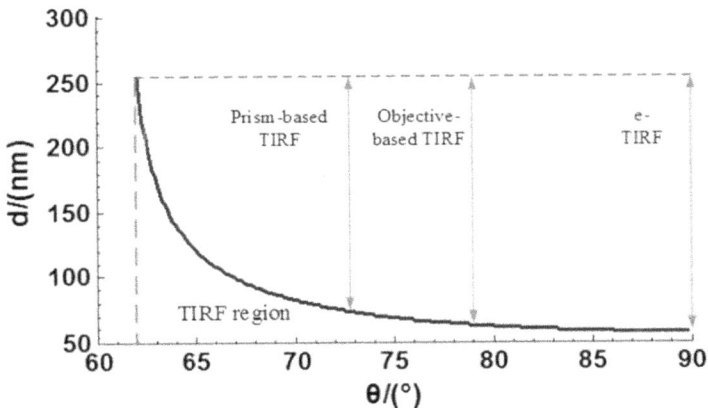

Figure 9.3. The relationship of incident angle and penetration depth in three TIRF methods [1].

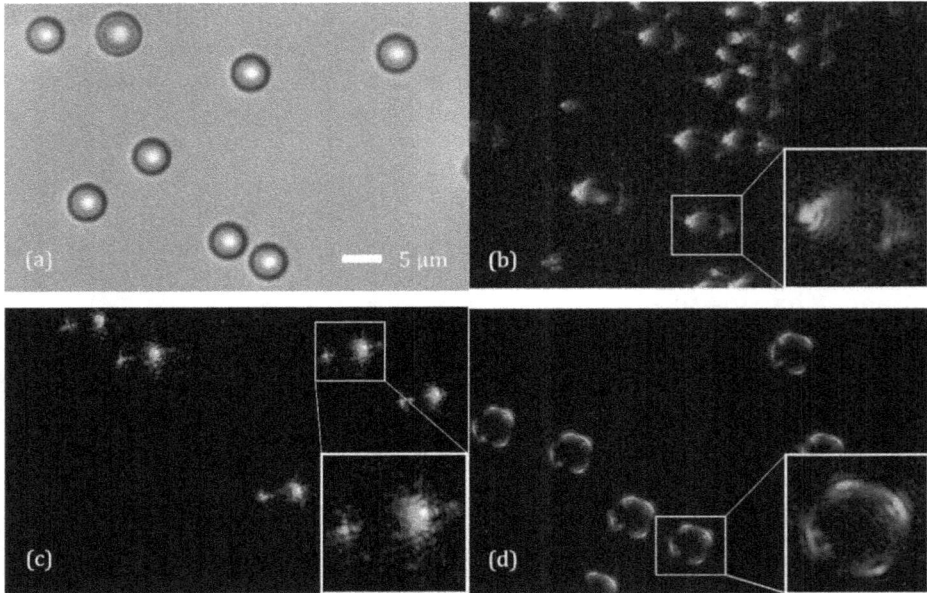

Figure 9.4. Images of wide filed with a fiber optic illuminator and three imaging modes with a laser [1]. (a) Wide field image. (b) Propagating light illumination image. (c) Prism-based TIRF image. (d) E-TIRF image.

method, a right angle prism is stuck to the glass slide with microscope immersion oil to direct light to the contiguous surface. Then we adjust incident angle to ensure that the incident angle is greater than the critical angle in figure 9.4(c). An elliptical mirror-based TIR image is shown in figure 9.4(d). The scale bar represents 5 μm in figure 9.4.

In figure 9.4(b), it can be obviously observed that there are several interference fringes on the downstream side of microspheres, with details in the enlarged region indicated by the white box. These irregular cometary fringes are caused by interference, because the local phase of coherent light is changed by optical imperfections in light path such as microspheres under the signal direction oblique illumination. In figure 9.4(c) the prism-based TIRF method, interference effects and background noise are lower than above because of TIR surface-selective illumination. However, there are still some beacons and shadows appearing next to the microspheres, which are created by the propagating light converted from the local evanescent field by the microspheres. In figure 9.4(d), we can observe microspheres clearly that shadow effects are attenuated and microspheres are well-distributed ring shape in this image. With the hollow cone beam illuminating in elliptical mirror-based TIR method, the same phenomena where evanescent light is converted into propagating light would be occurring, but the downstream effect is removed due to excitation from all azimuthal illumination directions. Although this demonstration does not eliminate ghost images, but makes them symmetric (or almost symmetric). According to the experimental result of an elliptical mirror-based TIR method, shadow effects are weakened by shadowless illumination.

In order to demonstrate fluorescent effects in e-TIRF microscope, we image fluorescence microspheres (lumisphere, 5 μm, silicon dioxide, fluorescent orange, base line) with prism-based TIRF method and e-TIRF method respectively. In figure 9.5(a), it can be observed that the orange part of fluorescence microspheres on the right is brighter than the left in prism-based TIRF method, with details in the enlarged region indicated by the white box in figure 9.5(a) region 1. By contrast, the orange brightness of fluorescent microspheres is symmetric in e-TIRF method in figure 9.5(b) and detail in region 1. Furthermore, for both sets of region 2 indicating the normalized strength of fluorescence microspheres, the red area is semicircular, and yellow and green region diffuses to left, and blue area is fuzzy in prism-based TIRF method. However, in e-TIRF method, the red region is circular at the center of microspheres, with yellow and green area surrounded. The blue area is more uniform than above. This experiment demonstrates that the intensity distribution of fluorescent microspheres is symmetric and has lower background noise in e-TIRF microscope which supplies shadowless illumination.

In order to demonstrate the ability of e-TIRF method to adjust penetration depth, we image fluorescence microspheres (lumisphere, 5 μm, silicon dioxide, fluorescent orange, base line) with different incident angle. In the experiments, if we only adjust the diaphragm, the external diameter of annular beam changes while the inner diameter of annular beam remains unchanged. Accordingly, for the hollow cone beam in e-TIRF method, the minimum incident angle (θ_{min}) relating to the external diameter varies and the maximum incident angle (θ_{max}) corresponding to the inner diameter is a fixed. The images obtained with different incident angle ranges are shown in figure 9.6. The allowable value of θ_{min} changes toward the critical angle.

In figure 9.6(a), there are several fluorescence microspheres appearing slightly though the whole image is still dim when θ_{min} is between 75° and 80° roughly. In figure 9.6(b), more fluorescence microspheres appear with distinguishable features. These microspheres become brighter with clearer contours, stronger contrasts when θ_{min} range is between 65° and 70° approximately. As shown in formula (1), the penetration depth is related to incidence light angle. When θ_{min} is between 75° and 80°, the penetration depth is so small that only the tiny part at the bottom of

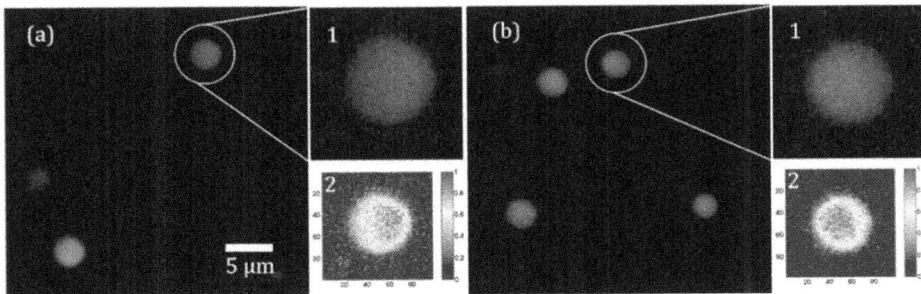

Figure 9.5. (a) The fluorescence microspheres images in prism-based TIRF method [1]. (b) The fluorescence microspheres images in e-TIRF method. For both sets of images, region 1 is the detail with enlarged scale. Region 2 indicates the normalized strength of fluorescence microspheres.

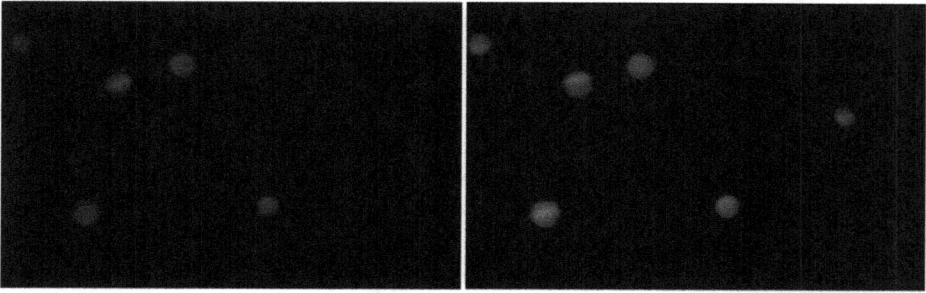

Figure 9.6. The fluorescence microspheres images at various θ_{min}, where θ_{min} is the minimum incident angle. (a) θ_{min} (75°, 80°). (b) θ_{min} (65°, 70°).

fluorescence microspheres are excited by evanescent field. With the penetration depth enlarging, the fluorescence microspheres become brighter and some new ones appear with greater distance from the contiguous interface. Therefore, the ability to adjust penetration depth is demonstrated in the e-TIRF method.

9.5 Summary

In conclusions, we propose an e-TIRF microscopy with a core elliptical mirror. (1) The elliptical mirror, which is similar to a shadowless lamp, supplies high efficient shadowless illumination with all azimuthal illumination directions and a large range of incident angle. It is experimentally supported by imaging the symmetric ring shape of microspheres, and the fluorescence microspheres images with symmetric intensity distribution and low background noise further demonstrate that shadow effects are attenuated. (2) The incident angle range can be controlled by adjusting the aperture size and the position of mask. Penetration depth is theoretically adjustable from 58 nm to 250 nm, whose range is wider than other two TIRF methods. (3) In addition, e-TIRF microscopy is of a lower cost because the fine high NA objective is replaced by an elliptical mirror.

E-TIRF enlarges the dynamic range of penetration depth and minimizes the limit value of minimum depth through applying an elliptical mirror, which is potentially useful for developing new techniques to acquire high axial resolution in the future.

References

[1] Liu J, Li Q, Li M, Gao S, Liu C, Zou L and Tan J 2017 Elliptical mirror-based TIRF microscopy with shadowless illumination and adjustable penetration depth *Opt. Lett.* **42** 2587–90
[2] Mattheyses A L, Simon S M and Rappoport J Z 2010 Imaging with total internal reflection fluorescence microscopy for the cell biologist *J. Cell Sci.* **123** 3621
[3] Boulanger J, Gueudry C, Münch D, Cinquin B, Paul-Gilloteaux P, Bardin S, Guérin C, Senger F, Blanchoin L and Salamero J 2014 Fast high-resolution 3D total internal reflection fluorescence microscopy by incidence angle scanning and azimuthal averaging *Proc. Natl. Acad. Sci* **111** 17164

[4] Pendharker S, Shende S, Newman W, Ogg S, Nazemifard N and Jacob Z 2016 Axial super-resolution evanescent wave tomography *Opt. Lett.* **41** 5499

[5] Axelrod D 2012 Fluorescence excitation and imaging of single molecules near dielectric-coated and bare surfaces: a theoretical study *J. Microsc* **247** 147

[6] Ma Y, Benda A, Nicovich P R and Gaus K 2016 Measuring membrane association and protein diffusion within membranes with supercritical angle fluorescence microscopy *Biomed. Opt. Express* **7** 1561

[7] Marsh G and Waugh R E 2015 A simple approach for bioactive surface calibration using evanescent waves *J. Microsc.* **262** 245

[8] Axelrod D 1981 Cell-substrate contacts illuminated by total internal reflection fluorescence *J. Cell Biol.* **89** 141

[9] Axelrod D 2008 Total internal reflection fluorescence microscopy *Methods Cell Biol* **89** 169

[10] Lin J and Hoppe A D 2013 Uniform total internal reflection fluorescence illumination enables live cell fluorescence resonance energy transfer microscopy *Microsc. Microanal.* **19** 350

[11] Fiolka R, Belyaev Y, Ewers H and Stemmer A 2008 Even illumination in total internal reflection fluorescence microscopy using laser light *Microsc. Res. Tech.* **71** 45

[12] Mattheyses A L, Shaw K and Axelrod D 2006 Effective elimination of laser interference fringing in fluorescence microscopy by spinning azimuthal incidence angle *Microsc. Res. Tech.* **69** 642

[13] Zong W, Huang X, Zhang C, Yuan T, Zhu L L, Fan M and Chen L 2014 Shadowless-illuminated variable-angle TIRF (siva-TIRF) microscopy for the observation of spatial-temporal dynamics in live cells *Biomed. Opt. Express* **5** 1530

[14] Lei M and Zumbusch A 2010 Total-internal-reflection fluorescence microscopy with W-shaped axicon mirrors *Opt. Lett.* **35** 4057

[15] Fu Y, Winter P W, Rojas R, Wang V, McAuliffe M and Patterson G H 2016 Axial superresolution via multiangle TIRF microscopy with sequential imaging and photobleaching *Proc. Natl. Acad. Sci* **113** 4368

[16] Stock K, Sailer R, Strauss W S L, Lyttek M, Steiner R and Schneckenburger H 2003 Variable-angle total internal reflection fluorescence microscopy (VA-TIRFM): realization and application of a compact illumination device *J. Microsc.* **211** 19